T0215261

HANDBOOK
OF
AERIAL MAPPING AND
PHOTOGRAMMETRY

HANDBOOK

OF

AERIAL MAPPING AND PHOTOGRAMMETRY

BY

LYLE G. TROREY, M.B.E.,

B.Sc. (LOND.), PH.D. (LOND.), M.E.I.C.,

A.R.P.S., P.ENG.,

*Member of the Canadian Institute of
Forestry*

*Member of the American Society of
Photogrammetry*

WITH A FOREWORD BY

W. W. WILLIAMS, *Lecturer in Surveying in the
University of Cambridge*

CAMBRIDGE

AT THE UNIVERSITY PRESS

1952

CAMBRIDGE UNIVERSITY PRESS
Cambridge, New York, Melbourne, Madrid, Cape Town,
Singapore, São Paulo, Delhi, Mexico City

Cambridge University Press
The Edinburgh Building, Cambridge CB2 8RU, UK

Published in the United States of America by Cambridge University Press, New York

www.cambridge.org
Information on this title: www.cambridge.org/9781107623309

First printed 1950
Second edition 1952
First published 1952
First paperback edition 2013

PUBLISHERS' NOTE
In this second edition the author has corrected
errors and misprints, has added notes referring
to latest knowledge, and has added a new
Appendix (III) which gives details of the Kelsh
plotter.

A catalogue record for this publication is available from the British Library

ISBN 978-1-107-62330-9 Paperback

TO
COLONEL H. L. MEUSER
O. B. E., R. C. E.,

Sometime, Director of Surveys,
Canadian Army Overseas

FOREWORD

The war of 1939-45 provides an interesting landmark in the progress of air surveys. In 1939 the science was young, but yet sufficiently advanced to leave no doubt that it had come to stay. Excellent maps had been made in Switzerland, Germany, Holland, Canada, the U.S.A. and, to a lesser extent, in other countries. While it was admitted that 'photogrammetry', as the science is now called, had possibilities, there was no general agreement as to its place in the survey world as a whole.

On the continent the instrument manufacturers saw new scope for their genius. The instruments of Zeiss in Germany, Wild in Switzerland, and Santoni in Italy, masterpieces of design and precision, were beginning to be used by government and by contracting surveyors, though not in this country. Here, interest was of a different kind. A few plotting instruments were designed in England, and experimental models were built, but the official solution of the air-survey problem was geometrical, not optical or mechanical. The treatment of tilted photographs was effected by mathematical analysis rather than by optical projection. This difference of treatment in 1939 is a most noticeable feature to the student of the history of photogrammetric development.

The literature of the subject up to this time tells the same story. Continental authors, dealing with the various problems in solid geometry, seek their solutions by instrumental methods, and their results call for the greatest admiration. The British literature, on the other hand, advocates graphical and mathematical solution, and, as far as I am aware, every method advocated requires an arduous treatment of the effects of tilt. The only instruments required are simple co-ordinate measuring devices which measure in the planes of the photographs, and not in what may be called 'terrestrial' co-ordinates.

Both are theoretically sound, though they had not then been so widely used as to convince the critics that they were serious survey methods. Many of the critics were hostile, or at least not progressive, and they would not admit the simple truth that in full-scale tests air surveys had yielded results which compared favourably with ground surveys in respect of accuracy. It can be assumed, then, that in 1939 photogrammetry was capable of obtaining a very desirable degree of accuracy—although this was not known to, or at least was not admitted by, many surveyors. Its survival, therefore, depended upon convincing map makers in general of this truth, and perhaps even more, upon the economy of air survey.

Surveys had to be paid for, and those who pay for them usually do so grudgingly. Costs, therefore, must be kept as low as possible, and costs in their turn depend very closely upon the rate of output. The war forced this issue. To fight the retreating Germans we needed maps of thousands of square miles, and air surveys offered the only means of producing them. Moreover, the time for this vast programme was short. If the maps were to be in time for the operations concerned they had to be prepared quickly. The pre-war tempo was too slow altogether. Thus photogrammetry was, in a sense, forced by the war into a position of prominence in the survey world.

It was at this stage that Trorey appeared. He served continuously during the war with Canadian Military Survey. Throughout his service he saw his task quite clearly. His object was to map very large areas of the continent in a very short time. His staff, when he arrived, was largely untrained, and at first he had little equipment. Major Trorey had had experience of air surveys in British Columbia, he had an open mind, and a great deal of determination. He was quick to try a method, and equally quick to discard it if proved unsatisfactory. Not all were destined simply to be accepted or rejected; some were modified, and eventually Major Trorey had 'a club for every lie', to use a golfing expression. For a method to be acceptable it had to be accurate, rapid, foolproof and simple.

The proof of the pudding is in the eating, and the results proved the success of the methods which Trorey either selected or evolved. He has no love of mathematical dalliance for its own sake, and will not tolerate unnecessarily complex devices. Common sense was the order of the day, and his inventiveness was directed to simplifying methods or adapting them to mass production rather than to academic research.

Dr Trorey has written a book for the man who wants to make maps from air photographs. It is a book for the drawing office, that is to say a manual.

He describes in detail methods which have been tried, and which have been found to work. Formulae are quoted, though not necessarily proved; those who are interested in the proofs can find them elsewhere. A large number of examples have been worked out in such careful detail that the application of the formulae is made very clear indeed. Towards the end of the book Dr Trorey writes: 'Although the metamorphosis from the blank paper to the finished three-colour sheet may seem complex and mysterious, the whole may be broken down into a number of unit operations, each of which is simple and easy to understand.' This, I think, is a true description of this book. Photogrammetry and stereoscopy can be difficult to comprehend, but Dr Trorey has here set out an admirably clear treatment of the whole problem by breaking it down into a number of small comprehensive processes, each of which is fully explained.

W. W. WILLIAMS
Lecturer in Surveying
Cambridge University

CAMBRIDGE
28 *April* 1947

PREFACE

If a map made for civil purposes contains inaccuracies or mistakes, inconvenience and serious financial loss may result. In a military map the consequences may be far graver—perhaps even to the extent of jeopardizing the whole operation. The responsibility of a maker of military maps is heavy. The map tells a story to those who read it, and the report it gives must be true.

In the actual making of the military map, the ground cannot be occupied either for control survey or for the amplification of photographic interpretation. The military cartographer is thus required to make a better map, and to make it from data less complete. He may be required to make it within limits of time quite unprecedented in civil practice, and he must do this with personnel who, for the most part, may be new to the work.

Many of the new and modified techniques described in this book were devised, in the first instance, to solve specific military mapping problems encountered in the North-West European Theatre. Nevertheless the text, as now presented, has primary application to civilian mapping. Some of the material was published during the war for restricted military circulation, particularly in 'Parallax Without Tears', 'Field Survey Pamphlet No. 6', 'Heighting Without Tears' and in 'Air Survey Research Paper No. 18'. Other new techniques developed and used during the war have not previously been published, either because it was felt, at the time, that they had not been sufficiently tried and proved against ground checks, or because there was no time to write the methods out in form suitable for publication.

Neither was there time to follow many promising lines of investigation—in multiplex operation; in connection with templet techniques such as the spring-loaded intersection locator and the error formula. With the end of the war this situation changed; it became possible to gather up many of these loose ends, and the relaxation of security regulations now permits publication.

The Chinese put the roof on a house before they build the rest, claiming, with some reason, that they are thereby able to carry on the work without interference from the weather. Photogrammetry is here presented in much the same manner, and for much the same reason. The first chapter deals, it is true, with the theory of perspective geometry—but only to provide the minimum amount of scaffolding to support the roof. The remainder of the theory is presented as it is required, usually in the form of exercises, or worked examples, of a particular technique.

War-time experience in making practical operational photogrammetrists from inexperienced personnel, against time, has shown the method to be workable. No attempt has been made to establish the more complex photogrammetric theorems which, in fact, are not in general used in the systems of air mapping described. Discussions and proofs of such theorems may be found in the works listed in the bibliography.

Von Gruber, Bagley, Hotine, Hart, Church, Finsterwalder and many others have dealt fully with the theory of photogrammetry, and with many of its applications. The justification for adding still another book must then be new methods based on old theorems, and simple, specific, working descriptions showing how these methods are applied in an actual air-mapping operation, together with reports of field tests showing the precision obtained in practice by the use of the techniques described. Research was needed to decide whether a particular new or old method was to be accepted, rejected, or modified. Most of this was done in the course of training successive classes of air-survey personnel, for a very important factor both in civil and in military air mapping is, how quickly can ordinary men be trained to operational proficiency in the use of this instrument or that method.

I particularly desire to acknowledge the help received in making these investigations, often under trying circumstances, from W. K. MacDonald, B. F. Engler, F. R. Dufton, H. W. Burton, M. P. Hollinger, J. I. Thompson, A. M. Nelson and the late E. F. Klemmer, all then of the Corps of Royal Canadian Engineers. Without these men the work could not have been carried to a successful conclusion.

I also desire to acknowledge the help and encouragement received from Professor Debenham of the Department of Geography at Cambridge, from Professor C. A. Hart, then of the Directorate of Military Survey, and from Major W. W. Williams, during the period in which he was on the staff of the Survey Directorate, R.E., and also, after the war, when he returned to the Department of Geography.

L. G. T.

LONDON
May 1947

CONTENTS

CHAPTER 1
PERSPECTIVE PRINCIPLES

Photogrammetry, 1. Anharmonic Ratios, 4. The Scheimpflug Condition, 6.

CHAPTER 2
MEASUREMENT OF ANGLES FROM OBLIQUES

Classification of Air-Survey Photographs, 7. Dip of the Horizon, 8. Azimuth and Depression Angles from Obliques, Graphical Method, 10. Analytical Method of Obtaining Azimuth and Depression, 16. Oblique Plotting Instruments: the Burns Plotter, 17. Azimuth Grid, 19. Construction of Rectangular Azimuth Grid, 19. Ground-Survey Photographs, 22.

CHAPTER 3
PERSPECTIVE GRID AND FOUR-POINT METHODS

Perspective Grids, 23. Proof of hv and hD Formulae, 25. Standard Grids, 26. Plotting of Forward Oblique Strips (Perspective Grid), 27. Projection of Perspective Grid on Plane containing Three Control Points, 29. Control, 30. Four-point Method, Drill, 31. The Use of the Four-point Method for Minor Control, 32. Shore-line Mapping, 34.

CHAPTER 4
MEASUREMENT OF HEIGHT FROM A SINGLE OBLIQUE

Height-Displacement Formulae, 37. Drill for Semi-graphic Height Determination, 37. Heighting Grids, 40. Determination of Height of Aircraft, and Measurement of Sloping Objects, by means of the Canadian Oblique Analyser, Mk. III, 49. Summary of the Uses of Obliques, 54.

CHAPTER 5
VERTICALS: FUNDAMENTAL CONSIDERATIONS OF COVER, PARALLAX AND STEREOSCOPY

Verticals, 56. Parallax, 58. Binocular Vision and the Stereoscope, 62. The Use of the Parallax Bar to determine Parallax Differences, 67.

CHAPTER 6
PARALLAX AND ELEVATION CALCULATIONS

Determination of Elevation Differences, Simple Case, 71. Standard Heighting Drill, Right Angles to Base-line, 72. Contouring, 74. Correction by Observation of K, 75. Parallax Scale, 81. Observational and other Errors and their Magnitude, 85. Survey Analogy to Heighting Problem, 85. Focal Length and Parallax, Model Effects, 87. Limitations of Photogrammetric Methods, 90.

APPENDIX 2
MULTIPLEX AND STEREOPLANIGRAPH: CONSIDERATIONS GOVERNING MINIMUM CONTOUR INTERVAL, 172.

TABLES INCLUDED IN THE TEXT

REFERENCES

These works are referred to in the text by number

(1) *Photogrammetry, Collected Lectures and Essays*. O. von Gruber. Chapman and Hall, Ltd., London, 1932.

(2) *Surveying from Air Photographs*. Air Survey Committee Professional Paper No. 8. Lieut. J. S. A. Salt. H.M. Stationery Office, London, 1939.

(3) *Use of Aerial Photographs for Mapping*. Topographical Survey (of Canada) Bulletin No. 62. The King's Printer, Ottawa, 1932.

(4) *Engineering Applications of Aerial and Terrestrial Photogrammetry*. B. B. Talley. Sir Isaac Pitman and Sons, Ltd., London, 1939.

(5) *Air Survey*. Second edition (corrected up to 1 November 1939). Pamphlet forming Chapter XII of the Survey of India *Handbook of Topography*. Major R. Crone, R.E., Superintendent, Survey of India. Geodetic Branch Office, Survey of India, Dehra Dun, 1939.

(6) *Surveying*. Vol. II. Higher. Fifth edition. Breed and Hosmer. Chapter XII, Aerial Surveying. John Wiley and Sons, Ltd., London and New York, 1938.

(7) *Surveying from Air Photographs*. Col. M. Hotine, R.E. Constable, London, 1931.

(8) 'Survey by High Obliques: The Canadian Plotter and Crone's Graphical Solution.' Lyle G. Trorey. *Journal of the Royal Geographical Society, London*, Vol. C, No. 2, August, 1942.

(9) *Air Photography Applied to Surveying*. Major C. A. Hart, R.E. Longmans Green and Co., London, 1943.

(10) *A Method of Determining Relative Tilt*. Lt.-Col. E. H. Thompson, R.E. The War Office, London, June 1944.

(11) *Tree Heights from Air Photographs by Simple Parallax Measurements*. G. S. Andrews. The Association of Professional Engineers in B.C. (Canada), 1936.

(12) *Student Reports of Three Research Problems in Aerial Photogrammetry*. No. 7. *Design of a New Stereoscopic Instrument for Topographic Mapping*. Muzzafer Tugal, Syracuse University, Syracuse, New York, 1943.

(13) *Aero-photography and Aero-surveying*. Lt.-Col. J. W. Bagley (Ret.), U.S.A. McGraw-Hill, New York, 1941.

(14) 'Methods of Surveying Laminaria Beds.' V. J. Chapman. *Journal of the Marine Biological Association, London*, Sept. 1944.

(15) *Tilt Correction to Height Measurements*. Major E. H. Thompson, R.E. Survey Directorate, G.H.Q., Middle East, Cairo, 1941.

(16) *Manual of Photogrammetry*. American Society of Photogrammetry, Pitman Publishing Corporation, New York, 1944.

(17) *Photogrammetric Engineering*. American Society of Photogrammetry, P.O. Box 18, Benjamin Franklin Station, Washington, D.C., Vol. XII, No. 2, June 1946.

(18) *Miscellaneous Publication*, No. 404. Harry T. Kelsh. United States Department of Agriculture, Washington, D.C.

(19) *Air Survey Research Paper*, No. 18. *A Multiplex Technique for Radar-controlled Air Photographs*. Major Lyle G. Trorey, M.B.E. War Office, London, December 1945.

NOTATION

Distortion is due to lens inaccuracies, paper and film shrinkage, and the like.

Displacement results from tilt or from relief or from both.

Points on the positive or negative image plane, or diapositive plane are represented by lower-case letters. Their homologues on the map, grid, ground, or stereoscopic model are represented by capitals:

f	Focal length.
S	Perspective centre, air station.
H	Height of perspective centre above datum.
h	Elevation of a point above datum.
h	(Of obliques) Intersection of principal vertical with true horizon.
h′	Intersection of principal vertical with visible horizon.
I, I	Isocentre.
N, n	Plumb point.
P, p	Principal point.
v	Vanishing point.
A	True horizontal angle subtended at N.
B	Depression angle, measured from the true horizon.
d	Depression of optical axis from the true horizon.
d′	Depression of optical axis from the visible horizon.
d″	Dip of the visible horizon.
θ	Angle between perspective planes, tilt.
ϕ	Angle of lens.
x, y	Co-ordinates of picture point, origin at p.
P	Parallax of an image point, millimetres.
p	Difference in parallax between two image points, millimetres.
c	Correction.
m	Datum point.
t	Tie point.
B	Actual air-base, S_1, S_2.
b	Photo air-base.
K	Want of correspondence, y parallax.

CHAPTER 1

PERSPECTIVE PRINCIPLES

Photogrammetry

PHOTOGRAMMETRY involves determination of the spacial co-ordinates of a point by means of two or more photographs containing it. Since in reality it is only angles which are measured photogrammetrically, the above definition implies the necessity of ground control, or the equivalent, for solution of the problem. Direct measurement of photographic co-ordinates or distances may or may not be involved—but when photographic measurements are made it is important to note that the purpose of such measurements is, directly or indirectly, to determine an angle. In special cases the problem may be solved by the use of one photograph only, and, in particular, angles may be measured from a single photograph. Since, however, most applications involve photographic triangulation, it is seen that, in the general case, at least two photographs from different viewpoints are required.

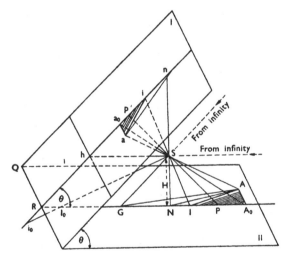

Fig. 1. Perspective diagram.

Since a photograph is a perspective projection, photogrammetry may be considered as inverse perspective. A photogrammetrist is required to use (inverse) perspective principles to make a map or a plan, or to obtain other specific information from air photographs. Perhaps the most important application of photogrammetry is in mapping from air photographs, and it is to this application that this handbook is, primarily, directed.*

---

* While the above definition is correct, the word photogrammetry is often commonly used to include the whole broad field of air survey operations and applications.

The more important perspective principles, having regard to a single perspective projection or photograph, are discussed and defined below. Fig. 1 is any bundle of perspective rays, centre S, cut by the planes I and II. These planes I and II may be the picture plane and the ground plane, the screen and the easel of the epidiascope, the two planes of a photographic rectifier or enlarger, and so on.

Definitions and Properties of Points of a Perspective Projection

PERSPECTIVE CENTRE.   S is the perspective centre, the centre of the lens. Neglect for the time being thick lens effects, internal and external perspective centres, nodal points and so forth. The geometry, slightly modified, applies, but it is simpler to consider a thin lens, or a pinhole, with optical centre S.

PLUMB LINE, PRINCIPAL AXIS.   Perpendiculars from S to I and II fall at p and N respectively. If II is the ground plane, N is the ground plumb point, and its projection on plane I is n, the photo plumb point, nSN being the plumb line. Similarly, pSP is the principal axis and p the principal point. The length of the perpendicular Sp is f, and that of SN is H.

PRINCIPAL PLANE.   The principal plane of the system is perpendicular to I and II and contains the perspective centre, S.   This is the plane pRN. Rp and RN are therefore at right angles to QR, and the angle pRN is θ.

ANGLE BETWEEN PERSPECTIVE PLANES. θ is the angle between these two planes. When θ is zero, the planes are parallel, and we have a vertical photograph. When θ is 90° the camera axis is horizontal, and the plate vertical—the usual condition for ground-survey photographs.

θ, as above defined, may be referred to as the tilt of the photograph, that is to say, θ is the amount, in degrees, of departure of the principal axis from true verticality. Tilt, and certain other terms, are sometimes used in a more specific sense. To avoid confusion, this and related terms will be defined now.

Referring to a single photograph, and without regard to its relation to any other adjacent photographs, tilt is θ of fig. 1. When θ becomes large it is customary to refer to the depression (from the horizontal) of the principal axis—that is to say, to $90° - \theta$. This is called the depression angle, d. Thus we refer to a tilt of 5° but to a depression angle of 25°, in which latter case θ is 65°. Note that rotation of the whole camera about the principal axis will have no effect upon the tilt.

Referring now to photographs which are nearly vertical, consider a single overlapping pair. θ of the first photograph may be resolved into two components mutually at right angles. It is usual to resolve θ into one component in the direction of flight, and one at right angles thereto. The axes of reference are the x-axis, joining the two perspective centres, the y-axis, which is horizontal and at right angles to the x-axis at S, and the z-axis, perpendicular from S to the plane containing the x- and y-axes.

The component in the direction of flight is often called fore-and-aft tilt. It is tilt *about* the y-axis. The term tip is commonly used, particularly in America, for this tilt. The component about the x-axis, lateral tilt, is often referred to simply as tilt.

Thus we have:

Rotation about the x-axis is lateral tilt, or simply 'tilt'. List is a synonym.

Rotation about the y-axis is fore-and-aft tilt, or tip.

The preferred American practice is to call these x-tilt and y-tilt respectively.

All of the above are absolute, that is to say, are measured from the horizontal. One also refers to *relative* tilt between two photographs—relative fore-and-aft tilt, relative lateral tilt or relative tip and tilt. In this work the expression 'tilt', unqualified, means the angle θ of fig. 1. Rotation about the z-axis is swing.

HORIZON LINES. Sh, in the principal plane, is parallel to plane II. A line on plane I through h, at right angles to Rh, will therefore be the projected position of the (true) horizon of plane II. This line through h is called the horizon line. Rays from all points infinitely distant from QR on plane II will be contained on a plane through S and parallel to plane II. All such points must, therefore, project on the horizon line. In addition, all parallel lines will converge to the same point on the horizon. These important properties of the horizon are the basis of many photogrammetric constructions. In the same way, SG is parallel to plane I, and a line at right angles to GN, in plane II, would be the horizon line for points infinitely distant on plane I.

ISOCENTRES. Sp and SN are the perpendiculars from S to planes I and II, as stated above. pSn is the internal angle between these perpendiculars, pSN the external angle. The bisectors of these angles are Si and SI_0 respectively. The points i and I_0 are called isocentres. They are also referred to as metapoles or conjugate focal points. Note that there are two isocentres, but that usually only one, that is, i, will appear on the picture plane.

PROPERTIES OF THE ISOCENTRE. The importance of the isocentre in photogrammetry is that angles measured on plane I from the isocentre i are the same as though measured from I on plane II. For example, the angle aia_0 on the picture plane is equal to the angle AIA_0 on plane II, provided that A and A_0 are both in plane II. That is, any photograph is angle true with respect to its isocentre for all points in the plane of the isocentre. It is *not* true if A and/or A_0 are displaced from plane II.

PROPERTIES OF THE PLUMB POINT. Since NSn is, by definition, vertical, the plane containing NASna is a vertical plane. Now if the point A be displaced vertically from plane II to some position A' (not shown in fig. 1), its new projected position in plane I, a' (also not shown), must be somewhere along na produced, since it is still contained in the vertical plane of NASna. We then have the following properties of the plumb point:

(i) Elevation displacement is radial from the plumb point.

(ii) If we set a row of picquets in line on the ground, either by eye or accurately with a theodolite, we habitually refer to this as a straight line. Of course it is not, particularly if the ground be irregular. It is the trace of a vertical plane with the ground. However, this conception of a straight survey line, of a straight road or railway, is so common that we must continue to use it.

Any such straight line on the ground—whether the terrain be level or undulating—if it pass through the ground plumb point will photograph as a straight line passing through the photo plumb point.

Set up a theodolite on the ground over the ground plumb point, and stake out any straight line NA. This line is the trace on the ground of the vertical plane in which the telescope transits. Imagine the photograph to be taken from a stationary helicopter, or from a balloon. Since you are set up vertically below S, if you transit the telescope the line of the sight will trace out a vertical plane passing through S, and thus contain n and a on the picture plane.

Other straight survey lines on the ground will only appear straight on the photograph in the special case that they are geometric straight lines. For example, a long railway tangent at a constant grade—it need not be level— will photograph as a straight line whether or not it passes through A. At a change of grade the line on the photograph will bend unless the tangent pass through N.

(iii) Conversely, all straight lines drawn on the photograph to pass through the photo plumb point contain detail which would lie on a surveyed straight line on the ground.

ISOCENTRE AND PLUMB POINT. Recapitulation: elevation displacement is always radial from the plumb point. Angles measured from the photo plumb point are not equal to the corresponding angles measured from the ground. *An oblique, or tilted, vertical is angle true with regard to its isocentres only if there is no ground relief.*

HOMOLOGOUS POINTS. Any point in plane II, such as A, has a corresponding projected position, a, in plane I. Such pairs as A and a are called homologous points.

Anharmonic Ratios

A PLANE IS DETERMINED BY S AND ANY PAIR OF POINTS. Three points determine a plane; thus any pair of points on a perspective plane and the perspective centre are contained by a plane. In fig. 1, the pair A_0I and S, determine the plane A_0IS, which plane, produced, cuts plane I in a_0i and QR in R.

So far we have considered the negative plane, plane I of fig. 1. For most purposes it will be simpler to consider a plane parallel to plane I at the same distance in front of the perspective centre as plane I is behind it; such a plane is called the positive plane. In fig. 14, p. 25, both the positive and the negative planes are shown. All the geometry of the negative plane will hold for the positive plane, since the image points on the latter bear the same relation to those on the former as does a contact print to the negative from which it is made. Hereafter, the positive plane is the plane referred to, unless otherwise stated.

INTERCEPTS OF THIS PLANE WITH THE POSITIVE PLANE AND THE GROUND PLANE. In fig. 2, a positive plane I intersects the ground plane II in QR. Imagine these planes hinged along QR, and let plane I be turned

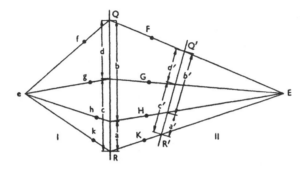

Fig. 2. Anharmonic ratios.

about QR until it becomes coplanar with II, now the plane of the paper. E, F, G, H and K are points in plane II, their corresponding projected positions in the positive plane I are e, f, g, h and k. The planes determined by S and EF, EG, EH and EK intersect plane II as before. The intercepts, on RQ, of the traces of these planes are a, b, c and d, as shown. Q'R' is any transversal whatsoever, either on I or on II, or on either set of rays produced —produced on either side of the pole.

PERSPECTIVE PROJECTION SETS UP ANHARMONIC RATIOS. It can be shown that a series of anharmonic ratios is set up by the perspective relations between object and image. In fig. 2 the ratio $a/b:c/d$ is an anharmonic ratio of these four variables. It can also be shown that $a/b:c/d=a'/b':c'/d'=k$, where k is a constant. The proof of the above relations can be obtained by successive application of the sine formula.

The foregoing is the basis of the four-point method of point transference which will be explained under near verticals.

Exercise 1. Draw an anharmonic diagram as fig. 2. Verify, by measurement of two transfersals such as Q'R', one on either side of QR, that the anharmonic ratios set up by these and by QR are the same.

The Scheimpflug Condition

APPLICATION. This is the condition of operation of epidiascopes, rectifiers and the like. The proposition was first stated by Abbe, but its practical significance was not recognized until much later when it was applied by Scheimpflug. When non-vertical photographs are to be rectified optically, to obtain sharpness of definition the condition must be satisfied except for small tilts.

THE CONDITION. The principal plane *of the lens* is the plane through its optical centre and normal to the optical axis. This should not be confused with the principal plane *of the projection*, namely, pRNS. The Scheimpflug condition is satisfied when the principal plane of the lens S of fig. 1 contains the line QR, the intersection of the perspective planes I and II. Then the lens S can project homologous points on these planes sharply in focus, and in the geometric relations shown, provided the ordinary optical condition of conjugate focal points is satisfied.

In oblique air photography, it is not necessary that the Scheimpflug condition be satisfied because, as all amateur photographers will know, for distances over about 100 ft. the lens may be focused for infinity. In air-survey cameras this hyperfocal distance, as it is called, varies from 100 to 900 ft. Even in obliques containing the horizon, the foreground is quite distant—accordingly, both it and the horizon will focus sharply with the optical axis set normal to the picture plane. In air-survey camera calibration, care is taken that the optical axis is normal to the focal plane within close limits.

CHAPTER 2

MEASUREMENT OF ANGLES FROM OBLIQUES

Classification of Air-Survey Photographs

GROUND-SURVEY PHOTOGRAPHS. These are usually taken with the optical axis accurately horizontal, and the plate is accordingly vertical. They are used infrequently in combination with air photography, but are valuable for extension of control. Their treatment is briefly discussed herein.

HIGH (OR HORIZON) OBLIQUES. Photographs taken with the optical axis tilted to include the visible horizon are called high obliques. The expression 'high' has nothing to do with the altitude of the aircraft; it refers to the camera being pointed high enough to include the horizon. Ordinarily when obliques are mentioned, it is high obliques which are thought of. Horizon obliques is a better designation.

LOW (NON-HORIZON) OBLIQUES. As the optical axis is further depressed from the horizontal the photo position of the horizon rises. When the horizon no longer appears on the format the photography is called low oblique. It is preferable to call such photographs non-horizon obliques. 'Verticals' too tilted to be dealt with by the radial line method may be called near verticals; they are seldom deliberately so taken for survey purposes but result from a variety of causes, the chief of which in wartime is the impracticability of maintaining standards of survey flying in the face of opposition from enemy aircraft and anti-aircraft batteries. In peacetime, too, such photographs may result in a variety of cases (e.g. in exploratory surveys). They are capable of utilization by various simple methods, and have a specific application in the determination of heights of ground objects (i.e. trees).

For military reconnaissance purposes a three-camera set-up is frequently used, the cameras being set at (approximately) 15° tilt to port, 15° to starboard, and vertical. This setting results in split verticals with tilts which may amount to 20 or 30°. Photography of this kind is sometimes the best that is available for military mapping. Split verticals are also used in forest inventory photography.

VERTICALS. The camera axis is vertical, or nearly so. Angle θ of fig. 1 is zero to 3 or 4°. It is stated that if tilts can be limited to 2° and the relief of the ground does not vary by more than 10% of the altitude of the flight[a], then the error so introduced is negligible for radial-line methods of plotting. In wartime survey flying it is to be expected that these limits may be exceeded. The treatment of such photographs will depend upon the magnitude of tilt actually encountered, the result required, the apparatus and the control available, and the skill and ingenuity of the photogrammetrist. Though tilts of 4°—with occasional photos of the strip exceeding this—are certainly undesirable if special photogrammetric apparatus is not available, such conditions usually can be handled best by the radial-line method, with which are included the slotted templet and similar devices.

Dip of the Horizon

TRUE AND APPARENT HORIZONS. Since the earth is a sphere, and not a plane surface, the horizon on a photograph will appear below the true horizon, the true horizon being that of the datum plane tangent to the earth's surface at N. (Strictly speaking, we should consider not the earth's surface, but an imaginary spherical surface at datum level.)

CALCULATION OF DIP IN MINUTES. In fig. 3, NU is tangential to the earth's surface at the plumb point, i.e. is horizontal. Sh is parallel to NU and accordingly is horizontal. ST is tangent to the earth's surface from S,

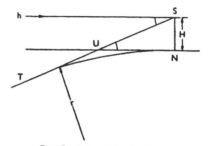

Fig. 3. Dip of the horizon.

at the point T. Dip of the horizon $= hSU = SUN$. $H = TS^2/2r$, where r is the radius of the earth. From the above $TS = \sqrt{(2rH)}$. But $TU = US$ (nearly) $= \frac{1}{2}TS = \frac{1}{2}\sqrt{(2rH)} = \sqrt{(\frac{1}{2}rH)}$.

Sin dip = dip in radians (since we are dealing with small angles); therefore

$$\text{dip in radians} = \sin \text{dip} = H/US,$$

$$H + \sqrt{\frac{rH}{2}} = \sqrt{\frac{2H}{r}}.$$

Substituting $r = 20.9 \times 10^6$ ft. and multiplying by $57.3 \times 60 \times 60$, dip in seconds

$$= 63.8 \sqrt{H}.$$

Taking the coefficient of refraction as 0.92, dip

$$= 58.5 \sqrt{H} \text{ sec.}$$

$$= \sqrt{H} \text{ min., nearly.}$$

At $f = 8$ in., $1°$ is of the order of $3\frac{1}{2}$ mm.; hence $1'$ of arc is about 0.06 mm. on the print. Accordingly, the $\sqrt{H}$ formula is of sufficient precision, and a slide rule is used for the calculations.

DIP OF THE HORIZON IN FEET. The dip of the horizon, in feet below the horizontal datum plane through N, must be known for certain purposes. This may be calculated from the well-known formula for the tangent offset to a circular curve; we have

$$c = D^2/2r,$$

where c is the correction for curvature alone, D is the distance from the ground plumb point to the point under consideration, and r is the radius of the earth. r, c and D are in the same units.

Certain published tables are based upon the refraction coefficient 0·86, the figure ordinarily used in ground survey. However, the index 0·92 seems to give better results at usual flight altitudes. Using the index 0·92 and substituting the mean radius of the earth in feet (20·9 million), the total correction C becomes

$$C = D^2/45 \cdot 4,$$

where D is the distance in thousands of feet.

CORRECTION TABLE (FT.)

TABLE 1. *Correction for curvature and refraction (thousands of feet)*

D (thousands of ft.)	C (ft.)	D (thousands of ft.)	C (ft.)
10	2	75	124
20	9	80	141
30	20	85	159
40	35	90	178
50	55	95	198
60	80	100	220
70	108	105	243

CORRECTION TABLE (METRES). Where D is in kilometres the correction in metres is

$$C = D^2/14 \cdot 9.$$

TABLE 2. *Correction of curvature and refraction (kilometres and metres)*

D (km.)	C (m.)	D (km.)	C (m.)
5	2	22	32
10	7	24	39
14	13	26	45
16	17	28	53
18	22	30	60
20	27	32	69

EXERCISES

Exercise 2. H = 3000 ft., f = 8 × 5 × 5 in. format. What is the inclination of the optical axis to ensure that the *true* horizon is just contained in the picture?

Referring to fig. 13 (*a*), Sh is horizontal, and h is the top of the photograph. Inclination of Sp just to contain the true horizon is $\frac{1}{2}\phi$, the half-angle of the lens, $\tan^{-1} ph/f = \tan^{-1} 2 \cdot 5/8 = 0 \cdot 3125$, whence the required inclination is 17° 21′.

Exercise 3. Data as problem 2. What is the inclination of the optical axis to ensure that the *visible* horizon is just included?

Referring to fig. 13 (*a*), owing to the effects of curvature and refraction the visible horizon is below the true horizon. The angle of depression, in minutes, has been shown to be equal to the square root of H in feet. For the data of this problem then,

$$d'', \text{ depression of horizon} = \sqrt{3000} = 55'.$$

The optical axis can therefore be depressed a further 55', i.e. to 18° 16', when the visible horizon will just be contained.

Azimuth and Depression Angles from Obliques, Graphical Method

AZIMUTH AND DEPRESSION. In survey and astronomy the azimuth of a point is defined as the arc of the horizon between the meridian and the vertical circle through the point(6). The azimuth of a point on an oblique is similarly defined as the arc of the horizon, or angle, between the principal plane and the vertical plane containing S, N, and the point. Azimuth of a vertical is, similarly, measured from the base-line.

The inclination to the horizontal of a ray from S to a point is the depression angle of that point. This is the angle which would be measured by the vertical circle of a theodolite, set up in space at S, levelled, and the telescope directed to the point.

For many purposes it is simpler to consider, not the depression angle as above, but the projection of this angle on the principal plane. Such principal plane angles are used in the Crone and other constructions. Consider a plane containing S and the point, and such that its trace on the photograph is a line parallel to the true horizon; the angle between this plane and the horizontal is the principal plane depression angle.

A nice application of obliques is in the measurement of azimuth and depression angles. By these means ground control may be extended with accuracy such that the position, and the elevation, of the image points so determined may be used to control vertical or other photography. A horizon, though desirable, is not necessary, as its position may be determined quite readily from ground control.

RELATION TO GROUND SURVEY. The methods used to determine azimuth and depression angles are photogrammetric, but, having determined these, further operations are identical to survey with which the reader is assumed to be familiar. For most practical purposes we may consider a series of theodolites at successive positions of the aircraft at exposure. The plate is levelled by the horizon, or by ground control, rather than by a bubble. In fact, everything can be done which could be done by these imaginary theodolites, the only restriction being that they cannot be swung outside the field of the camera.

There will be described the following methods of determining azimuth and depression angles: graphical method, analytical method, instrumental methods, azimuth grid.

GRAPHICAL (CRONE) CONSTRUCTION FOR AZIMUTH ANGLES. A rapid and ingenious method of taking off azimuth (the Crone method) has been developed by the Survey of India(5). It is, of course, applicable *irrespective of the ground relief.* Drill is as follows, fig. 4 representing the photograph and fig. 5 a sheet of Kodatrace.

Photograph

Kodatrace

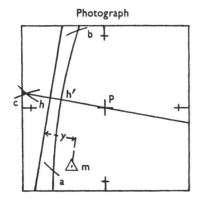
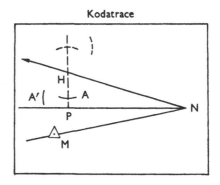

Fig. 4. Photograph. Crone construction for azimuth.

Fig. 5. Azimuth fan. Crone construction for azimuth.

(i) On the photograph establish, if possible, the true horizon, and draw the principal vertical. Construction (fig. 4). With centre p strike an arc to cut the visible horizon at a and b, near the edges of the photograph. With centre a and suitable radius strike an arc c, and with centre b and the same radius strike an arc to intersect the first arc as shown. Join p to this intersection, establishing the principal vertical. (Fig. 4 is purposely drawn with a large swing, and with the curvature of the horizon exaggerated.) The true horizon is above the apparent horizon by a distance which can be calculated as in Exercise 3. If the visible horizon is not clearly defined, draw a trial horizon from the best information available and draw the trial principal vertical.

(ii) On a piece of Kodatrace draw PN, making PN = f (fig. 5). Through P draw PH at right angles to PN, making PH = ph of the photograph. Join NH, *distinguishing this line with an arrowhead.*

(iii) It is required to locate a ray from the ground plumb point N, passing through the point M. Set dividers to dimension y of fig. 4, the distance of m below the true or trial horizon of the picture. With centre H and radius y strike an arc to intersect HP, or HP produced, as shown in fig. 5. With centre at this intersection and the same radius, strike a second arc A' *on the side of* HP *remote from* N.

(iv) Place the trace on the photograph with NH, the line marked with the arrowhead, along the principal vertical and, maintaining this coincidence, slide the trace along ph until the arc A' is tangent to the horizon. Prick the position of the image point m at M. Join NM on the trace. Then NM is the required ray and the angle HNM is the azimuth of M.

(v) In mountainous country, the true horizon may be below the image points of the higher peaks, as when their elevation is greater than the height of the aircraft. In these instances the sign of y changes, and the arcs A and A' of fig. 5 will be drawn *above* HN and to the *right* of HP, as shown dotted in fig. 5.

RESECTION OF PLUMB POINT. Three control points Q_1, Q_2, Q_3 such as M, being contained on one photograph, the ray from N to each of these may be obtained, and N resected by the ordinary solution of the three-point problem, the trace (fig. 5) being superimposed on the map or control grid. This locates the points Q_1, Q_2, Q_3, N and the principal line NR (fig. 6) at the scale of the grid or map. Note that a large-scale vertical photograph may be substituted for the map, provided h/H be small.

GRAPHICAL METHOD OF ESTABLISHING TRUE HORIZON. Where the true horizon does not appear on the photograph, or in the circumstances of (v) above, or where it is desired to establish the horizon rather accurately, the same authority gives the following solution. The construction also serves to determine H–h, and elevations, as closely as they may be scaled.

On non-horizon obliques the true horizon, even though it be several inches off the print, may be required. If three points of ground control appear in such a photograph, the Crone construction may be used to establish the true horizon, after which it may be treated in all respects as a horizon oblique. Whenever the Crone construction is used, f should be known as accurately as one can draw the line Sp, say plus or minus $\frac{2}{10}$ mm. Working on paper prints, an equivalent focal length should be calculated from measurement of the distance between the fiducial marks on the print.

Choose a trial horizon on the photograph and resect the plumb point as described above (figs. 5 and 6). This trial horizon is put on from the best information available, and it will generally be possible to locate it within a quarter of an inch or so. The plumb point having been resected, the drill is as follows:

(i) On fig. 6 draw Q_1R_1, Q_2R_2, Q_3R_3 at right angles to the principal line.

(ii) On fig. 7 draw pS = f, scale full size, and from p draw the photoplane at right angles to pS. Set dividers to the distance ph (fig. 7a), obtaining h. Join Sh produced. On fig. 7 mark $SR_1 = NR_1$, etc., of fig. 6, and from R_1, R_2, R_3 drop perpendiculars.

(iii) On the photograph set dividers to the y ordinate (as fig. 4) of each point in succession, transferring them to fig. 7, at y_1, y_2 and y_3.

(iv) Rays from S through y_1, y_2, y_3 will now locate Q_1, Q_2, Q_3 on the principal plane in the position shown.

(v) The elevation of Q_1 is known. From it must be *subtracted* the correction for curvature and refraction in feet (or metres) (tables 1 and 2, p. 9). It is possible that, in flattish country, the corrected elevation may be negative. Let the corrected elevation be Q_1T_1. Mark T_1 as shown. Similarly with Q_2 and Q_3.

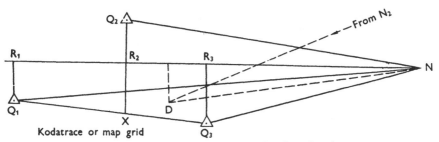

Fig. 6. Resection. Crone construction for azimuth.

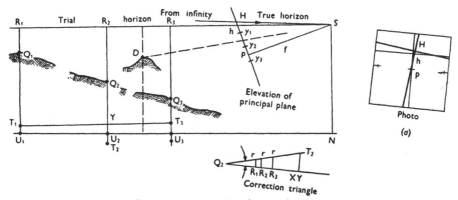

Fig. 7. Crone construction for true horizon.

(vi) If the trial horizon is correct, these points will lie in a straight line parallel to Sh. If the three points lie in a straight line which is not parallel to Sh, through S draw SH parallel to T_1, T_2, T_3, and establish the true horizon on the photo parallel to, and at a distance hH from, the trial horizon. The usual case, however, is where the trial horizon is displaced from, and at an angle to, the true horizon. Then the points T will not lie in a straight line, and the procedure will be as under:

(vii) On fig. 7 join T_1T_3, cutting R_2T_2, produced if necessary, in Q. In fig. 6 join Q_1Q_3 (the same corresponding points as T_1T_3), cutting Q_2R_2, produced if necessary, in X.

(viii) Lay off the correction triangle of base equal to Q_2X and height T_2Y, the angle at X, Y being a right angle. On the base mark R_1, R_2, R_3, making Q_2R_1, Q_2R_2, Q_2R_3 equal respectively to Q_1R_1, Q_2R_2, Q_3R_3 of fig. 6. Some of these points R may fall to the right of X, Y. From each R erect a perpendicular to cut the hypotenuse in r as shown.

(ix) From T_1 mark $T_1U_1 = Rr_1$ from the correction diagram. The direction in which T_1U_1 is to be marked is the same as the direction Q_1 to R_1 of fig. 6, when TY is in the same direction as Q_2X_2. Where this direction is *opposite*, as in the case illustrated, mark the TU direction *oppositely* to the respective QR directions.

(x) Join U_1, U_2 and U_3 which should now be co-lineal. Through S, draw SH parallel thereto, establishing the true horizon in elevation. Drop the perpendicular SN, representing the true height of the air station above datum.

(xi) To transfer the true horizon to the photograph, on the trial principal vertical mark hH and draw the true horizon inclined to the trial horizon by an amount equal to the angle at Q in the correction triangle. The *left* side of the true horizon will be *above* the trial horizon if T (fig. 7) for a point on the *left* of the photo is *above* U, and vice versa, as at (*a*) (fig. 7). In fig. 7, Q_3 is on the left of the photo, and T_3 is above U_3. Hence the angle at Q in the correction triangle will be set off on the photo so that the left side of the true horizon is above the trial horizon, as shown.

(xii) If the correction, particularly of the angle at Q, has been great, repeat the resection of fig. 6 to ensure that N remains unchanged.

CUTTING IN ELEVATIONS. As further points such as D are established in plan by intersection from two N's, their elevations above datum can be determined graphically by scaling the vertical distance from D to the datum. Correction for curvature and refraction must be *added* to the scaled height.

CRONE CONSTRUCTION FOR TRUE HORIZON; MODIFICATION FOR EXAGGERATED VERTICAL SCALE. In certain circumstances it would be convenient to work the foregoing construction at an exaggerated vertical scale, similar to the ordinary procedure used in plotting a road or railway profile.

The following modification effects this, enabling certain parts of the construction to be carried out with greater precision, and rather more easily. It is emphasized, however, that the modified method requires accurate work, especially in erecting the perpendicular through p. If the work is careless, errors introduced in the graphic multiplication may outweigh the gain obtained by plotting TU, and the correction triangle, at the enlarged vertical scale.

Enlargement in any degree may be obtained by a construction similar to that described, but two, or at the utmost three times, seems to be the useful limit.

DRILL FOR MODIFIED CONSTRUCTION (fig. 8).

(i) Draw pSH as for the standard construction of fig. 7.

(ii) From p drop an accurate perpendicular to Sh. The strength of the whole construction is dependent upon this right angle, which, therefore, must be drawn with precision. (1) With centre p and radius about 2 × ph strike arcs cutting Sh. (2) With centres at these two cuts, and with radius about twice that used in (1), strike two arcs to intersect as shown. (3) Join this cut to p, intersecting Sh in O, and produce past p as shown.

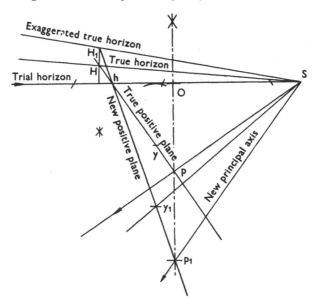

Fig. 8. Modified Crone construction for true horizon, enlarged vertical scale.

(iii) Let required enlargement of the vertical scale be two times. Set dividers to pO, and with centre p cut p_1 on the perpendicular, then p_1O equals 2 × pO. Join Sp_1 and hp_1, produced as shown. Sp_1 is the axis and hp_1 produced the positive plane for enlarged vertical scale.

(iv) To locate any ray on the new positive plane, proceed as follows. (1) On the true positive plane locate y as in the ordinary construction. (2) With centre y and radius the perpendicular distance to Sh, strike an arc intersecting the new positive plane in y_1. Then Sy_1 is the required ray.

(v) Repeat for each of the control points, and continue the construction exactly as previously laid down to operation (x) inclusive. Remember, when locating T, that we are now working at an enlarged vertical scale.

Having obtained the adjusted datum line:

(vi) Through S draw SH_1 parallel to the adjusted datum line intersecting the new positive plane in H_1.

(vii) From H_1 drop a perpendicular to Sh cutting the true positive plane in H, which point H is the same as the point H of the standard construction. Note that H_1H should be one-half the perpendicular distance from H_1 to Sh.

(viii) The correction triangle is drawn as before, but the angle TQxy is not the true correction angle for the horizon, as was the case in the previous construction, but is an angle whose tangent is twice that of the required angle. Bisect Ty in T′, then T′Qxy is the required angle.

Analytical Method of Obtaining Azimuth and Depression

AZIMUTH. The azimuth angle A, the angle MNH of fig. 5, may be calculated from the photo co-ordinates of the point m. In measuring photo co-ordinates the origin will be taken as the principal point p, and the usual sign convention is followed. That is to say, x is positive to the right and negative to the left of the principal vertical. The x-axis is an imaginary line through p at right angles to the principal vertical (parallel to the horizon). y is positive above this line, negative below it. A better degree of accuracy in measuring y is obtained by measuring the distance from the point to the true horizon, and subtracting the (known) distance ph, rather than by drawing the x-axis through p, and scaling from that.

Simple geometry gives the following:

$$\tan A = \frac{x}{f \cos d - y \sin d}.$$

There is little advantage in calculating tan A, since the accuracy depends upon how closely the co-ordinates, and ph, are measured. A seeming precision will be obtained in the answer, which precision is not justified by the accuracy with which x and y are measurable by ordinary means. Where accurate means are available for measuring photographic co-ordinates—a comparator, for example—and for determining paper distortion, then the calculation is justified.

Otherwise the Crone construction will actually draw the angle as closely as m may be pricked on the kodatrace. Further discussion of relative accuracies of various methods will be found in (8).

DEPRESSION. Referring to fig. 7, the depression angle B to any point D may be calculated from photo co-ordinates in a similar manner. This is the Crone depression angle, the angle measured in the principal plane:

$$B = HSQ' = d - y'Sp = d - \tan^{-1} y/f.$$

Again there seems little justification for the calculation. B may be measured by the tangent method when it is desired to calculate elevations, rather than to scale them off fig. 7.

EFFECT OF INCORRECT HORIZONS. Incorrectly placed horizons introduce errors in the angles measured by any of the above photogrammetric means, but, as we are usually concerned with angular *differences*, this is not quite so important as it might seem at first sight. This may be better understood by the following comparison. Consider a theodolite set up at the perspective centre; imagine this on a captive balloon if you like. Sight the instrument on any point on the principal vertical on the ground, and have the instrument set so that one plate bubble is parallel to the line of sight, and the other at right angles. If the plate is not level about a line at right angles·to the line of sight it will show up on the striding bubble; this is tilt of the horizon. If the plate is tilted along the line of sight as well it will show up on the other bubble; this is displacement of the horizon. Now the errors in measuring either horizontal or vertical angular differences with this incorrectly levelled theodolite are those you would get off a picture whose horizon is incorrectly located by the same amount.

This will explain how quite good differences in elevation, and quite good resections, may be obtained from horizons which are both displaced and tilted.

Oblique Plotting Instruments: the Burns Plotter

GENERAL CHARACTERISTICS. There are a number of instruments which consist, in effect, of a theodolite or alidade placed at the perspective centre of an oriented oblique. The line of sight being directed at a point of detail, its azimuth and depression may be determined. The Wilson photo-alidade, the Miller plotter, and the Burns plotter are examples. The Burns plotter is interesting in that it will plot planimetric detail directly, and without calculation, from a single oblique of flat country. However, its preferred use is in determining azimuth and depression. This it will do rapidly and with rather remarkable precision[8]. The instrument is described in detail below.

MECHANISM OF THE BURNS PLOTTER. Fig. 9 shows the machine in use and fig. 10 shows, diagrammatically, its construction in elevation of the principal plane. The photograph is placed upon the plate ph, its principal

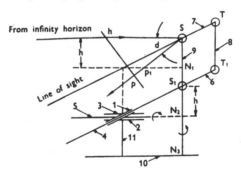

Fig. 10. Diagram of Burns plotter.

point coincident with the centre. The instrument is collimated so that the principal axis is normal to the plate ph when the telescope is directed at its central point. Thus when hSp is set at the true depression angle, and the plate swung about an axis through S until its centre registers with p, the orientation of the photograph is completed by simple rotation of the plate about p. Sp is adjustable for different focal lengths, and may be set accurately to the equivalent focal length of the print being used. Consider orientation as above to have been carried out, then planimetric detail may be taken off as follows:

ST is a telescope pivoted at S, and having a vertical circle at S and a horizontal circle on the SN axis. It is evident that if the telescope be directed at any image point, the line of sight will intersect the ground at the actual object photographed. Imagine an auxiliary plane N_1p_1 as shown. The trace of the line of sight on this plane would be a point homologous to the ground point at the scale, SN_1 divided by height of aircraft. 6–7–8–9 is a parallel motion; hence the space rod T_1S_1 produced (6, 4) will always be parallel to the line of sight, 7, TS produced. The space rod thus duplicates the motion of the line of sight as the telescope is made to scan the picture, and the trace of the space rod with plane N_2 is a point homologous with the trace of the line of sight with plane N_1, the effect of the parallel motion being the movement of the auxiliary plane from N_1 to N_2.

The members 1 and 2 are pivoted at 3, the space road slides in 1, while 5 is a long worm to which 2 is threaded. Rotation of 5 then moves 2, and the whole parallel motion, rotating the telescope in a vertical plane. Motion in azimuth results from the rotation of SN_3 by means of gearing not shown, and a horizontal circle at N_2 measures the rotation. These motions are transferred to the drawing table 10 by the pencil 11. Thus, as the telescope scans a piece of detail, a true plan, scale h/H, is traced on 10, provided no ground relief exists.

Measurement of Azimuth and Depression. The preferred application of the machine is to determine azimuth and depression angles in rough terrain. The picture being oriented as explained, and the telescope TS directed at an image point, azimuth and depression are read on the horizontal and vertical circles. Setting pS to an equivalent focal length to compensate paper distortion, angles may be turned off as by a similar theodolite on the ground.

Drill for Setting Up

(i) Determine equivalent focal length by measurement of the fiducial marks and set pS to this value. When non-uniform distortion exists, use the x equivalent for depression, and the y equivalent for azimuth. (This expedient is not mathematically precise, since x and y enter into both angles.)

(ii) Draw the principal vertical on the print and determine the depression of the optical axis, as previously described, or if necessary use the Crone construction.

(iii) Set the vertical circle to read d. Raise or lower the photo plate and adjust the azimuth handwheel until the cross-hairs intersect the central punch mark on the plate.

(iv) Put the photograph on the plate with the principal point on the cross-hair intersection, alining the principal vertical along the vertical hair.

(v) Swing the line of sight towards the top of the photo, and if the intersection of the cross-hairs does not now cut the principal vertical, rotate the plate by means of the plate tangent screw until it does. Swing to p again as a check. The set-up is now complete, and the azimuth and depression of any image point may be read directly off the circles.

Fig. 9. Burns high oblique plotter in use. (By permission of the Royal Geographic Society.)

To face p. 18

Caution. In (iv) and (v) care must be taken not to disturb, by accident, the azimuth handwheel setting obtained in (iii).

DRAWING AZIMUTH LINES. When a calculated resection is not desired, TS is directed at the image of a control point. Rotation of 5 moves the telescope in a vertical plane containing the plumb line SN_3; hence 11 moves radially to the plumb point, actually drawing a ray from the plumb point N_3, giving the true horizontal angle in plan.

Rather than with a horizon obtained as described above, a more nearly precise result is obtained from a horizon calculated, by a series of approximations, from ground control. Under these conditions trigonometric elevations may be thrown forward considerable distances; in fact, the results seem comparable to ground work with a minute reading theodolite [8].

Azimuth Grid

AZIMUTH LINES. A vertical plane containing the plumb line will intersect the ground in a straight line through the ground plumb point and project on the photograph as a straight line radial from the photo plumb point. A series of such planes, say $1°$ apart, would then result in ordinates on the print from which the angle subtended at the plumb point, between any two image points, could be read directly. The distance, along the horizon left or right of the principal vertical, of the trace of the plane at A degrees in azimuth therefrom, is f sec d tan A. The line is radial from the photo plumb point.

EQUIDEPRESSION LINES. Lines of equal depression angles, equidepression lines, are the intersections with the photo-plane of a series of cones, i.e. a series of hyperbolas of which the vertices lie in the principal vertical of the photograph. The distance from the horizon to the hyperbola for depression angle B, measured down the principal vertical, is ph − f tan (d − B).

AZIMUTH GRID. A grid composed of such azimuth and equidepression lines will serve to determine, by inspection, the azimuth and depression of any image point on the photograph. Angles are determinate on a 5×5 in. format within 5 or 6'. Such a grid is known as an azimuth grid [6], although this name may be somewhat misleading, as its characteristic is the hyperbolic depression line.

Construction of Rectangular Azimuth Grid

DEPRESSION ANGLES. For the determination of depression angles the conventional azimuth grid uses equidepression lines which, as explained, are hyperbolas formed by the intersections with the photo-plane of vertical cones containing the plumb line. These are angles such as would be read by a theodolite at the perspective centre at the instant of exposure, or by the Burns plotter. Crone in his graphical construction does not use the actual depression angle itself, but its *projection* on the principal plane, as shown in fig. 7. A grid made up to read Crone depression angles will then show depression lines as straight lines parallel to the horizon. The calculations and procedure in drawing a family of hyperbolas, or circular curves as approximations thereto, are involved even though tables have been constructed which reduce the labour—but it is obviously simpler to draw a straight line. Such a straight-line grid is called a rectangular azimuth grid, and is constructed as follows.

DRILL FOR CONSTRUCTION OF RECTANGULAR DEPRESSION LINES.

(i) Establish the true horizon and principal vertical, calculate d, the depression of the optical axis. Data required, f and an *approximation* to H, which latter is used only to determine the dip of the horizon.

(ii) Let there be a plane through S, the trace of which plane on the picture plane is a straight line parallel to the horizon, and let b be the intersection of the said line with the principal vertical. Then, where B is the inclination of this plane to the horizontal, hb = ph − f tan (d − B). Calculate hb for values of B from o to $\frac{1}{2}\phi$ − d, where ϕ is the angle of the lens. These calculations may be made in 5° increments of B, and units obtained by subdivision of the 5° spaces.

(iii) Having marked the values of hb on the principal vertical, and made the unit subdivisions, through these points draw a series of lines parallel to the horizon, accentuating the 5° lines.

(iv) In calculating elevations from these projected depression angles, the same procedure is followed as in the Crone method, that is to say, the lineal distance used is not the distance from N to the point in question, but its projected distance on the principal line in plan, i.e. the NR distance of fig. 11, and *not* the NQ distance.

AZIMUTH LINES. A vertical plane containing the plumb line will intersect the photo-plane in a straight line passing through the photo plumb point. Detail on the photo contained in this trace will lie in a straight line on the ground. Let ha be the distance from the principal vertical, measured along the horizon, of a plane at A° to the principal plane. Then by simple geometry ha = f sec d tan A.

DRILL FOR CONSTRUCTION OF AZIMUTH LINES:

(i) Calculate ha by 5° increments of A for the range of the picture, using the above formula. (Maximum value of A will be in excess of $\frac{1}{2}\phi$.)

(ii) The azimuth lines are radial from the photo plumb point n. n lies at a distance ph/f cot d from h. For f = 8 in. and d = 15° it is of the order of 32 in. Rather than locating n, mark n′ at a distance $\frac{1}{5}$hn from h. Through n′ draw a base-line parallel to the horizon.

(iii) On the horizon mark the calculated values of ha, and on the base-line mark values of n′a′, where n′a′ = 0·8 × ha. Join the successive aa′ points, establishing 5° azimuth lines. Units are obtainable by subdivision of the 5° spaces; the 5° lines should be accentuated as for depression.

Note. The figure $\frac{1}{5}$hn of (ii) above is chosen quite arbitrarily. Any suitable simple fraction can be chosen. With wide-angle lenses the plumb point may fall close to the photo, and in these circumstances n itself may be used. In this case merely join the successive positions of a to n.

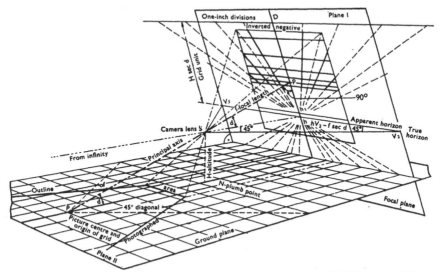

Fig. 11. Perspective grid principle. After a diagram by Gerard H. Matthes in (6).
(By permission of the publishers, John Wiley and Sons, Ltd., London and New York.)

ORDERS OF ACCURACY OBTAINABLE IN AZIMUTH MEASUREMENT. The orders of accuracy obtainable in practice, by each of these various means, is given in table 3. The table is based on actual results, but note that *good* photographs and *good* horizons are used and that paper distortion, if sensible, must be considered. See also (8).

TABLE 3. *Precision of azimuth determinations from oblique photographs*

Method	Accuracy (plus or minus) azimuth angles	Remarks
High oblique plotter	15″	Good photographs, very well-defined image points—horizon from ground control, calibrated f
Crone construction	1′	Same photographs and same points—experienced personnel
Crone construction	2′	Same, but by personnel in training
Analytical	2′	Same, photo co-ordinates scaled to nearest tenth of a millimetre
Grids	4′	Angles by inspection, compared to high oblique plotter

POSITION ACCURACY, INTERSECTIONS. Assuming the plumb point resection is strong, and that photographs are available from which an intersection angle of about 40° may be obtained, points may be intersected with probable position errors as shown in table 4.

In applying table 4 it is stressed that these results will only be attained *under the conditions stated* with reference to table 3.

TABLE 4

Distance from plumb point		Position error, metres for angular error of		
km.	miles	15″	1′	2′
10	6	2	9	17
20	12	4	17	34
30	19	6	25	51
40	25	9	34	68

Ground-Survey Photographs

ORDINARY CONDITION. The ordinary condition, as is implied from the definition of p. 2, is that d is zero and that the position and orientation of the horizon are known with precision. Where ground photographs are taken with d unequal to zero, treat as the equivalent oblique.

AZIMUTH AND DEPRESSION:

(i) Where $d = 0$, graphical construction is as follows: Draw the photo-plane at right angles to a line pN of length f, and set off the x ordinate of the image point on the photo-plane, joining it to N, so obtaining its azimuth. This is merely a plan view of the lens and the dispositive plane.

(ii) Depression may be obtained in a manner similar to fig. 7, no correction for horizon being required, of course.

(iii) The analytical formulae given on p. 16 reduce to $\tan A = x/f$ and $\tan B = y/f$, B being the depression angle as projected on the principal plane. Or $\tan B = y/f \sec A$, where B is the true depression angle, as would be read by the high oblique plotter.

GROUND PHOTOGRAPHY A SPECIAL TYPE OF OBLIQUE. Ground photographs are thus briefly discussed, not because they are of little consequence, but because they are a special type of oblique, the treatment of which the exact location of the horizon and true verticality of the plate make rather simpler. No mention is made of the many ingenious stereo-photogrammetric machines for ground photography—and for obliques as well—since such instruments are not now in general use.

CHAPTER 3

PERSPECTIVE GRID AND FOUR-POINT METHODS

Perspective Grids

WHERE METHOD IS APPLICABLE. The use of perspective grids is primarily suited to country having little relief, and requires that the depression angle and height of aircraft be determinate. Referring to fig. 11 the method consists of constructing a perspective grid, such as would be produced by photographing a rectangular grid which might be imagined to exist on the ground. Detail on the photograph is transferred to the map grid, square by square. In practice, transparencies of a number of grids are made for a particular lens to cover the range of height and depression angle required.

DRILL FOR GRID CONSTRUCTION. The method of drawing up these grids will be best understood by working an example. Data: f=8 in., H=3000 ft. Given an oblique photograph, calculate and draw the proper perspective grid, the size of the grid squares being 100 m. (figs. 12, 13). The proof of the formulae of (iv) and (v) is given after the conclusion of the drill.

(i) Draw the principal vertical, i.e. the line through p at right angles to the apparent horizon. Construction (fig. 13). With centre p strike an arc to intersect the visible horizon at a and b, near the edges of the photograph. With centre a and suitable radius strike an arc c; with centre b and the same radius strike the arc to intersect the first arc as shown. Join p to this intersection, giving the principal vertical.

Notes. Fig. 13 is purposely drawn with a large swing, and with the curvature of the horizon exaggerated. It is well to mount the photograph on a sheet of litho stock.

(ii) Calculate angle d', the depression of the optical axis from the apparent horizon. In the example ph' measured 1·47 in.

Tan pSH' = 1·47/8 = 0·18375, psh' = 10° 25'
Dip of horizon = $\sqrt{3000}$, calculated as in Exercise 3 ____55'____
Therefore pSh = d = 11° 20'

(iii) Calculate hh', distance between true and apparent horizons (fig. 13(a)):   hh' = ph − ph' = f tan d − ph'

$$= 8 \times \tan 11° 20' - 1·47 = 0·13 \text{ in.}$$

Mark h, draw the true horizon through h parallel to the apparent horizon. Alternatively the grid may be drawn from a Crone horizon, which would be treated as a true horizon.

(iv) All lines on the photograph parallel to ph vanish at h. We wish to draw the perspective projection of a series of such lines 100 m. (328 ft.) apart; these will vanish at h. The distance hD, in inches, is equal to H sec d/grid unit in feet. Through D draw a line parallel to the true horizon. On

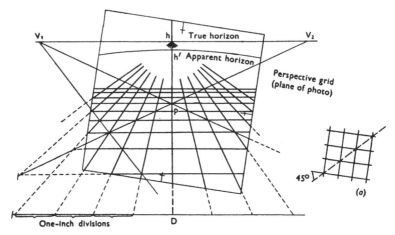

Fig. 12. Perspective grid, plane of photograph.

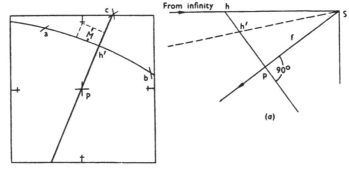

Fig. 13. Horizon and principal vertical.

it mark off divisions of 1 in., as shown in fig. 12; join these points to h. The resultant fan is the projection of a series of lines 100 m. apart on the ground, and parallel to the principal line NP:

$$hD = 3000 \times \sec 11° \ 20'/328 = 9\cdot33 \text{ in.}$$

(v) See (a), fig. 12. If a series of equally spaced parallels be cut by a line at 45°, and lines at right angles to these parallels be drawn from the inter-sections with diagonals as shown, the result is a series of squares. Now all lines at 45° to ph vanish at the same point; if we can calculate this point we can draw the projection of the diagonal, and so construct the projection of the squares. The distance in question is $hv_1 = hv_2$, of figs. 11 and 12:

$$hv_1 = f \sec d = 8\cdot00 \sec 11° \ 21' = 8\cdot16 \text{ in.}$$

Mark the point v_1 and draw v_1p produced. Now a series of lines at right angles to the principal vertical, through the points intersected by the 45° line pv_1, gives the required grid. v_1 and v_2 should both be used, to provide a check on draughting.

(vi) It is not necessary to carry the grid much past the principal point because the squares become so foreshortened that plotting from the background of the photograph is inaccurate. Accordingly, forward obliques are flown with rather more than 50 % overlap, to ensure that the area near p of the first picture will appear in the foreground of the second picture.

In dividing the line through D into 1 in. intervals, it is not necessary to carry the divisions out very far. The divisions for the upper part of the grid may be obtained by extending, on any parallel farther up the picture, the few equal divisions which will have been obtained on it by rays from D. These additional points are joined to h in the same way as were the divisions from D. This point will be clear when an actual grid is drawn.

Similarly, remembering that *all* 45° lines parallel to pv_1 vanish at v_1, an additional diagonal may be drawn to v_1 from the corner of any grid square. This diagonal will locate the parallels for the bottom of the picture.

The only reason for drawing the diagonal through the principal point in the first place is to make p come at the corner of a grid square.

Proof of hv and hD Formulae

SCALE OF A VERTICAL PHOTOGRAPH.   Referring to fig. 14, scale = ab/AB. In the triangles SNA, Sna, SNA = Sna = 90°; the angle at S is common, hence the triangles are similar and Sa/SA = Sn/SN.

As well the triangles SAB, Sab are similar, whence ab/AB = Sa/SA, which latter has just been shown to be equal to Sn/SN, i.e. to f/H.

Therefore scale = ab/AB = f/H. Thus the scale of a truly vertical photograph of *flat* country is the same all over the picture.

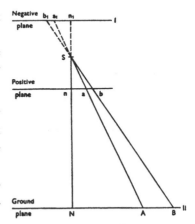

Fig. 14. Scale of a vertical photograph.

SCALE OF AN OBLIQUE.   Consider any plane containing S, such plane being at right angles to the principal plane. The trace of all such planes is a straight line, parallel to the horizon and at right angles to the principal vertical. In true plan such a plane would appear as fig. 14, where na would be the trace of such a plane with the positive photo-plane, and NA the trace with the (horizontal) ground plane. From similar geometry it is then seen that the scale of an oblique of flat country is constant on all lines at right angles to the principal vertical.

CALCULATION OF hD. In fig. 15 the scale at the point $p = Sp/SP = f \sin d/12H$, where H is in feet and f in inches.

On fig. 15 the scale at D is arbitrarily made 1 in. = grid unit. That is, the scale at D is $\frac{1}{12}$G, where the grid unit is in feet.

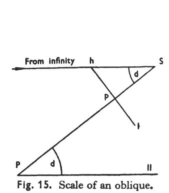

Fig. 15. Scale of an oblique.

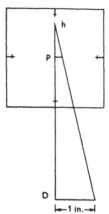

Fig. 16. Scale at point D.

Again from fig. 16 scale at D/scale at $p = hD/ph$. But $ph = f \tan hSp = f \tan d$. Substituting these values, and the known scale at D, in the above we have

$$hD = ph \times \text{scale at D/scale at } p$$
$$= f \sin d/\cos d \times \tfrac{1}{12}G \times 12H/f \sin d$$
$$= H \sec d/G.$$

Note that H and G are in the same units, and that hD is in inches.

POSITION OF VANISHING POINTS OF LINES AT 45° TO THE PRINCIPAL VERTICAL. It is required to find the position of the vanishing points, on the true horizon of the picture, of the projections of lines on the ground at 45° to PN of fig. 11. Let these points be at v. Then referring to the figure, vS is in a horizontal plane, parallel to the ground, and the angle $hSv = 45°$.

In the triangle hSv, since the angle at S is 45°, and the angle at p 90°, then the angle at v is 45°. Hence the triangle is isosceles, and $Sh = hv$.

In the triangle Sph, the angle at p is 90°, hence $Sh = Sp \sec f = \sec d$.

Therefore $hv = f \sec d$.

Standard Grids

NEED FOR STANDARD GRIDS. Having drawn a grid on a photograph, it will be realized that this method would prove rather slow and cumbersome in practice. Grids covering the range of height and depression angle likely to be encountered are drawn up for the particular lens being used. These are drawn three or four times full size, photo-reduced and transparencies made.

CANADIAN PRACTICE. Canadian practice as given in (3) is as follows:

(i) Marginal distances, rather than depression angles, are considered. Marginal distance is the distance M of fig. 13. Grids are made in $\frac{1}{10}$ in. increments of M covering the probable range of depression angles.

(ii) Altitude intervals are 50 ft., corresponding roughly to the $\frac{1}{10}$ in. marginal increment. The ordinary flying height in Canadian practice is 5000 ft.

(iii) The distance from the ground position of the principal point to the ground plumb point is worked out for each grid. This is PN of fig. 11, and PN = H cot d.

(iv) The foregoing information and the focal length of the camera lens are marked on each grid.

RELATION BETWEEN H, M AND SCALE. For H = 5000 ft., f = 8 in., d = 20°, a change in marginal distance of $\frac{1}{10}$ in. corresponds to a change in depression angle of about 38′. The maximum error in depression angle resulting from the use of grids calculated for $\frac{1}{10}$ in. increments in marginal distance would then be of the order of 20′. 10,000 ft. measured on the principal vertical, 5000 ft. either side of p, would change in length (as measured on the grid) about 40 ft. if a grid were used calculated for 20° 20′ instead of 20° 00′. This error is $\frac{1}{100}$ in. at 1 : 63,360, which is scarcely plottable. An equal difference would be introduced by using a grid calculated for a height 25 ft. in error. Thus 50 ft. increments in height are consistent with marginal distances in $\frac{1}{10}$ in. increments.

Plotting of Forward Oblique Strips (Perspective Grid)

AZIMUTH LINE. Obliques taken with the camera axis pointing in the direction of flight are known as forward obliques. On forward obliques azimuth can be carried forward simply by drawing a straight line on the first photograph, and producing it through successive overlaps by passing it through corresponding detail. This may be called an azimuth line.

If, throughout the strip, the flight has been straight, the azimuth line could be made to pass through the image positions of successive plumb points and would, in fact, be the principal vertical. In these circumstances, elevation displacement being radial from the plumb point, ground relief would not affect the accuracy of azimuth.

The flight line will not be accurately straight in practice, but the azimuth line will be kept as close as possible to the principal vertical in order that relief will affect it only to a small degree.

Lay out the flight in its approximate orientation, in such a way that an azimuth line may be chosen on the first picture which, when produced throughout the flight, will deviate as little as possible from the principal verticals of succeeding pictures.

The best line so obtainable may deviate too far from the principal verticals, a condition which will occur when the line of flight is a flat curve, or is very irregular. In these circumstances, a second azimuth line may be chosen parallel to the first one, remembering that parallel lines vanish at the same point on the horizon. Its distance from the initial azimuth is obtained from the grid. It is usually better to start the azimuth from the middle of the flight, working both ways.

CALCULATION OF H FROM CONTROL. On a photograph in which appear two ground control points—A and B of fig. 17—choose a grid of the correct marginal distance (or the correct ph') and for the altimeter height less the approximate elevation of the ground. The distance between these controls obtained from a preliminary plot of the grid will agree with their known distance apart if H-h is correct. If they differ, choose a new grid having calculated a corrected elevation as follows:

$$\frac{\text{Correct H}}{\text{H used}} = \frac{\text{True ground distance}}{\text{Ground distance from trial grid}}.$$

Exercise 4. From a trial grid, H = 3000 ft., the distance between control points is known to be 1126·4 m. apart, scaled 1210 m. What was the actual height of the aircraft above the ground, and what grid should be used?

$$\frac{\text{True height}}{3000} = \frac{1126·4}{1210},$$

$$\text{True height} = \frac{1126·4}{1210} \times 3000 = 2790 \text{ ft.}$$

The height is then 2790 ft. The nearest standard grid to this would be used, the 2800 ft. grid.

Note that it is *not* necessary to work this figure out to more than three significant figures (use a slide rule), since the answer is required only to the nearest 50 ft.

Exercise 5. In the above question, what error is introduced in the base-line AB by using the 2800 ft. grid, instead of a grid calculated for 2790 ft.? What will this error amount to in millimetres on scale 1:50,000?

$$\frac{\text{True distance}}{\text{Distance from 2800 grid}} = \frac{\text{True height}}{\text{Height used}},$$

$$\text{Distance on 2800 grid} = \frac{\text{Height used} \times \text{true distance}}{\text{True height}}$$

$$= \tfrac{2800}{2790} \times 1126·4 = 1130·4 \text{ m.}$$

Error introduced, 1130·4 − 1126·4 = 4·0 m.

More simply, error $= \tfrac{10}{2790} \times 1126 = 4·0$ m.

Scale 1:50,000, 4·0 m. $= \dfrac{4·0 \times 1000}{50,000} = 0·08$ mm.

SCALE CONTROL POINTS. Two points are chosen in the foreground of the second picture, and identified on the first picture. Such points are 1 and 2, fig. 17. The correct grid having been determined for picture 1, we may treat points 1 and 2 as ground control, so establishing the true height and correct grid for picture 2, in the same way that points A and B were used on the first picture. The process is repeated until the next ground control is reached, CD

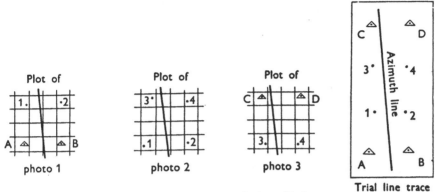

Fig. 17. Assembly, perspective grid plots.

of fig. 17. It is unlikely that the distance CD, as determined by the height carried from picture 1, will agree with the known distance from C to D, and an adjustment will be necessary. This may be referred to as internal adjustment.

Assembly Drill:

(i) From each photograph make a plot on paper gridded to correspond with the map grid. These plots show ground and scale control points, and azimuth (see fig. 17).

(ii) On a piece of tracing paper large enough to contain the flight, plot the ground control from the map grid—ABCD. Set this trial line trace on the plot of photo 1 by means of A and B, mark 1 and 2 and the azimuth line. 1 and 2 and the azimuth line on the trace are now placed on plot 2, whence 3 and 4 are obtained. Repeat until the next ground control CD is reached.

(iii) Adjustment. AB on the trial line will agree with AB on the map grid. The distance CD will also agree, having been made to agree in the internal adjustment. The distance AC (and the distance BD) will probably *not* agree. The grids on the first and last photos, having been fitted to ground control, will not be changed. The grids on the intervening photos will be changed to make AC on the plot agree with AC on the map grid.

Projection of Perspective Grid on Plane containing Three Control Points

hv Adjustment. The standard adjustment described fits a grid to two control points on a horizontal plane. This adjustment, alteration of the height for which the grid is calculated, will adjust a single plot accurately to two control points—but will not in general do more, since the change in height really amounts simply to a scale change.

It will often be found that two points are on within the limits we require, while the third—which may be in the middle background—is off. We therefore need to change the scale of the perspective grid in one direction only—not in both directions, as will result from the straight scale change of the height adjustment. This can be done by altering hv, hD remaining constant. The new hv is calculated as follows:

$$\frac{\text{True distance}}{\text{Distance from trial plot}} = \frac{\text{New hv}}{\text{hv used}}.$$

This will give a change of scale parallel to the principal vertical, the scale at right angles remaining unchanged.

HORIZON OF PLANE THROUGH THREE CONTROL POINTS. Suppose a photograph to have three control points at different elevations, and let us go through the Crone construction for true horizon, without, however, putting on the elevations (QT, fig. 7) for each control point. The horizon which the construction gives is not now that of a horizontal plane, but the horizon of a plane containing the three control points. The height given by the construction will be the distance of the perspective centre from this plane, not from the horizontal plane.

We may now go ahead and draw our perspective grid for this height and depression angle, and we thus project a grid on this plane. Taking off the positions of the controls on a trial grid as before, all three points will be very nearly correct. Further slight adjustment may be required because, though we have drawn a true perspective on the plane, the plane is not horizontal, and the map is the projection on the horizontal plane. We have measured slope distances and not horizontal distances.

APPLICATION TO MORE IRREGULAR COUNTRY. It will be seen that the use of a plane containing all three control points permits the extension of perspective grid methods to terrain otherwise unsuitable. When the control points may be sited, they will be chosen so that the plane containing them lies as close as possible to the main features it is desired to map. It is the 'elevation' of the remaining features above or below this non-horizontal datum plane which is the limiting factor. Thus for any given standard of mapping, the elevation range which can be tolerated is quite considerably increased.

It will be seen that circumstances may arise when a combination of the two adjustments may be required—but with a well-drawn Crone the adjustment will be so small that it is frequently unnecessary to make it at all.

Control

GROUND CONTROL AND MINOR CONTROL. It will now be understood what is meant by control. Points of which the position (and often the elevation) have been determined by ground survey are called ground-control points.

Even in wartime, precise (?) information is available as to the co-ordinates of many triangulation points in enemy territory.

By photogrammetric means, some of which will now be understood, the gaps between these ground-control points may be bridged and intermediate points established. These photogrammetric points—such as the points 1, 2, 3, 4 of fig. 17—are minor control points. Some kind of minor control plot, as it is called, is usually run starting from one, or from a group, of ground control points and tied in to other ground control at or near the end of the strip.

Our 'ground control' may be points established from special control pictures, such as horizon obliques, from which a number of strong, well-identified points are located—or may be position of the aircraft as determined by radar.

THE USE OF MINOR CONTROL. Near verticals have been defined—we could put on the horizon by the Crone construction and draw a perspective grid, treating such pictures as horizon obliques. We do not because it is scarcely practical to do so—with a tilt of 10°, depression angle 80°, for $f = 8$ in. the horizon would be 45·4 in. from the principal point.

To put on a Crone horizon, we would have to know the position of three photo-points and the focal length of the camera. If, on near vertical photographs, we know the map positions of four points, and if the country is rather flat, or—which is really what is meant—the relief is small compared to the height of the aircraft, we can rectify these tilted photographs, make them as they would have been had they really been verticals. The scale of a vertical of flat country is the same all over the picture, and we can quite easily make maps from such rectifications.

So we need to find the ground positions of four (or more) photo-points on each picture before we can make a map.

(*Note.* A photograph can be rectified knowing three points and the focal length of the cone.)

Four-point Method, Drill

FOUR-POINT METHOD. The four-point method can be applied to any photograph, vertical or oblique, *where the relief is small in comparison to* H. Though applicable to all photography fulfilling the above conditions, it is chiefly used on near verticals. When such a near vertical contains four control points, additional points may be established by this method, and if the points so established are chosen on the overlap, the control may be extended over a series of photographs. The method also serves to establish a network of triangles on the photograph and on the map-grid. Detail can be transferred by means of these triangles in a manner similar to that by which detail is taken from a perspective grid.

DRILL, FOUR-POINT METHOD. In fig. 18, I represents the photograph as plane I of fig. 2, and II represents the map, or ground plane II of fig. 2.

The positions of E, F, G and K are known, their corresponding photo positions are e, f, g and k. It is required to locate on II and point H, whose picture position is h. Drill is as follows:

(i) From e draw rays to f, g, h and k on the picture.

(ii) On the map grid on which E, F, G and K have been plotted, from E draw rays to F, G and K.

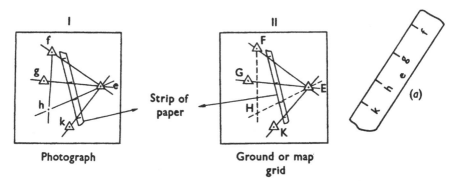

Fig. 18. Four-point construction. Four-point construction for true horizon.

(iii) Place a strip of paper on the photograph to cut each of the rays from e, as shown in fig. 18, I. Tick off the ray intersections on the strip of paper or cut with a razor blade. In lieu of a strip of paper, a piece of Koda-trace may be used. A line is scribed, and the points pricked.

(iv) Transfer the strip of paper to the map grid and superimpose it as shown, in such a position that the rays from E to the three other points cut the strip at the positions previously ticked from the photograph. Prick the position of the fourth ray.

(v) Since we have set up the same anharmonic ratios on I and II, h must lie on EH.

(vi) Repeat the above operations from f, obtaining the ray FH whose intersection with EH establishes the point H.

(vii) As a check, ray also from G, or, if desired, from K.

The Use of the Four-point Method for Minor Control

COMPARISON WITH RADIAL-LINE METHODS. The radial-line method (q.v.) is exact for an untilted vertical, no matter what relief exists. Where there is 10% ground relief and, at the same time, up to about 4° of tilt, corrections can be made, and it, or one of its modifications, is probably the preferred treatment. The four-point method has diametrically opposite properties, giving correct results over flat country for any photograph irrespective of tilt, from true vertical to high oblique. Where tilts up to 30° are present, it is known that the method can be used at 1:12,500 with 2% ground relief. For photography at 30,000 ft. it can be relied upon for planimetry with relief up to 600 ft. in combination with any tilts likely to be encountered in practice. It will not give satisfactory results for large tilts over deeply accidented country photographed from ordinary heights—unless primary and minor control may be sited at differences in elevation which are within the limits mentioned. The construction re-establishes a set of perspective conditions for one photograph, and carries this through the plot. It is essential that four accurate control points appear on one photograph.

APPLICATION TO RANDOM OVERLAPPING PHOTOGRAPHS. The technique is particularly suitable where random photography is the best available. Knowledge of f and H is not required, and 60% overlap is unnecessary for planimetry. There need only be sufficient overlap to permit four common points to be identified on adjacent photographs. This small overlap may be of photographs at different scales, taken with different cameras, at different times, from different heights, and with different tilts. These factors are not related to the accuracy of the extension of control, except in so far as scale is concerned.

RECTIFICATION NOT ESSENTIAL. The technique is suitable for use when no rectification means are available. The only equipment required, for planimetry, is a straight edge and a few scrap cuttings from the guillotine. The construction results in a network of triangles on the photo and on the grid. From these, detail can be transferred in the ordinary way—if the triangles are too large they may be readily broken down to any desired size.

STEREOSCOPY. While planimetry may be obtained from photographs without 60 % overlap, and on different scales, this observation does not apply to topography. The parallax bar cannot be used on near verticals with success, but form lines may be drawn by the usual methods if overlapping pairs on about the same scale be available. From sketchy vertical control, the drainage pattern, and *topographic sense*, quite good form lines can be produced. (Certain instruments can blend photographs at different scales and tilts in stereoscopic fusion, e.g. the Seely Duoscope.

DRAWING. It is advisable to work on the backs of the photographs, pricking the points through. The only caution here is that the pricker be a fine needle and be held vertical. No confusion will result from the reversal if the strip of paper be marked as fig. 18 a. Where the construction lines are required for transference of detail as above, the lines forming the required triangles will be transferred to the front of the photographs. Meanwhile, the fronts have been left clear, which makes stereoscopy and interpretation much easier.

The paper strip should be placed as far as possible from the pole being used, for obvious reasons. When control is being extended far it is advisable to fasten the print, upside down, to a sheet of drawing paper, and to extend the rays beyond the limits of the print. Starting from *four* control points, any triangle which appears comes from one cause, and from one cause only—inaccurate drawing. This is a pleasing feature of the method, since one knows immediately when a mistake or an inaccuracy has occurred. In the same way that three points establish a plane, four points establish one set, and only one set, of perspective conditions. In teaching the method, the above important proposition may be brought out by the following exercise.

Exercise 6.   Take two pieces of drawing paper, and on one, which we will consider as being the map grid, choose and mark at random four points, A, B, C and D. On another piece of drawing paper, which we assume to be a tilted photograph of these points, mark, also at random, a, b, c and d. Show that from any pole perfect intersections are obtained on the grid from the position of E.

ACCURACY OBTAINED.   Tests indicate that, for the conditions of about 2% relief at $H = 30,000$ ft., and with large tilts, control may be run from four trigs to a fifth 6 km. distant, with closing errors of the order of 100 m. After adjustment it is possible that the residual error at the centre of the plot is of the order of 20 m., assuming no side overlap with which to tie. This error is about 1 mm. at 1:25,000. The method is cumbersome, and requires patient work by a careful man. But it does provide a means of utilizing photography which, otherwise, could not be used for mapping.

Shore-line Mapping

FOUR-POINT HORIZON.   Since the high-water mark is a horizontal line, perspective grid methods are ideally suited for shore-line mapping. It often happens that, with available cameras, when you get the shore where you want it in the picture, the horizon is off the format. We may locate the horizon by the following means.

For the specific case where there are four control points at the same elevation the following direct graphical solution is available. It requires four control points, rather than the three necessary for Crone, but a knowledge of f, and of the position of ṗ, are unnecessary. The four points of ground control must, however, be at the same elevation, such as on a shore-line. Time required is about 15 min., exclusive of identification of control. A large-scale vertical photograph of the shore can be used in lieu of ground control.

DRILL FOR FOUR-POINT HORIZON.   Drill for graphical location of true horizon, having chosen and identified four control points, 1, 2, 3 and 4, on the map (or on the vertical) and on the oblique, is as follows:

(i) From photo-point 1 draw rays to each of the remaining controls (fig. 19). Draw also rays V_1–1 and V_2–1 such as would cut the left and right sides of the horizon respectively. These should be produced beyond 1 as shown.

(ii) On the map—or grid—ray the same corresponding points, and transfer V_1–1 and V_2–1 to the grid as in the four-point method.

(iii) On the grid ray from any other pole, 4, to each of the remaining controls (fig. 20). *Now, if* V_1 *and* V_2 *are on the true horizon they must, on the map, lie at an infinite distance from the pole* 4. Therefore the grid rays V_1–1 and V_1–4 must be parallel. Draw V_1–4 parallel to V_1–1. Similarly V_2–4 is drawn parallel to V_2–1.

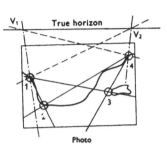

Fig. 19. Photograph. Four-point construction for true horizon.

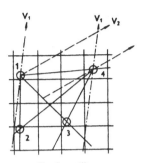

Fig. 20. Grid, map, or large-scale vertical photo. Four-point construction for true horizon.

(iv) Returning to the photograph, ray from 4 to each of the other controls, and transfer V_1–4 and V_2–4 from the grid. The intersections establish the true horizon V_2–V_1 as shown. (*Note.* Excepting for instructional purposes the rays 4–2, 4–3 will be drawn as a part of operation (i).)

(v) As a check on the accuracy of the work, ray from a third pole. Even a very small triangle at the trisection should be investigated and corrected.

NECESSITY FOR PRECISE WORK. The utmost care and precision must be taken with this construction, and in particular the point identification must be meticulously accurate. Where it is possible to choose the location of control points, they should not be sited in the foreground of the picture. The geometric strength of the figure does not appear to be very important, and in any case weakness shows up as a triangle in the trisection, and usually can be overcome by accurate draughtsmanship.

CHOICE OF HORIZON RAYS. Referring to figs. 19 and 20, rather than the rays to V_1 and V_2 as shown, rays such as 2–1 produced should be used where possible. This facilitates, and at the same time improves the accuracy of, operation (ii), as the step of transferring the anharmonic ratio is cut out.

CURVATURE AND REFRACTION. Since we are dealing with points at comparatively small distances from N, not at infinity, the effects of curvature and refraction are generally negligible.

CHECKS. It will be clearly understood by those familiar with the four-point method that, just as there is no check in making a resection from only three points, there is no check with only four points in the foregoing construction. Thus a control point may be wrongly identified on the print, or wrongly plotted on the grid, without the error showing up. Triangles at the trisection are caused by draughting only. A check is, however, provided when *five* points are used, as a blunder will then show up immediately the attempt is made to transfer anharmonic ratios.

FOUR-POINT HORIZON WHEN CONTROL POINTS ARE NOT AT THE SAME ELEVATION. Where the four points of control are not on the same plane, the height displacement may be calculated, set off, and the above construction carried out as described. The procedure used is as follows. Draw a trial horizon, as in Crone. Draw the right-hand portion of fig. 7, so that d and B may be determined; it is sufficient to measure these angles with a protractor. Calculate the height displacement D using the following formula:

$$D = \frac{fh}{H} \times \frac{\cos B \sin B}{\cos^2 (B-d)}.$$

Semi-graphic solution of this equation is quite suitable (see pp. 37 et seq.).

Locate the photo plumb point n (pn = f cot d) and join the control point under consideration to n. Set off D *parallel to the trial principal vertical.* From the extremity of this D line, drop a perpendicular to the trial principal vertical, the intersection of this perpendicular with the ray to n giving the undisplaced position of the control point.

It should be noted that three points may be reduced to the plane of the fourth, in which case H should be decreased by the elevation of the fourth point. Altimeter readings may be used to determine H, which is actually height of aircraft above the feature.

CHAPTER 4

MEASUREMENT OF HEIGHT FROM A SINGLE OBLIQUE

Height-Displacement Formulae

RIGID FORMULA. It can be shown that the height of a feature, h, is given by the following formula:

$$h = \frac{DH \cos(B_t - d) \cos(B_g = d)}{f \cos B_t \sin B_g}.$$

D is the height of the photographic image of the feature (measured parallel to the principal vertical), B_t and B_g are the depression angles to the top and to the bottom of the feature (measured in the principal plane), d is the depression of the optical axis, H the height of the aircraft above the (bottom of the) feature, and f the focal length of the cone.

APPROXIMATION. This equation (which is exact) may be solved by measurement of the picture co-ordinates of the top and bottom of the feature, calculation of the various angles and substitution. But analytical procedure is cumbersome and laborious, and the exact equation will reduce to the following without error commensurate with the method

$$h = \frac{DH \cos^2(B - d)}{f \cos B \sin B}.$$

B is here the depression angle to a point m midway between the top and the bottom of the feature.

This is capable of simple semi-graphic solution reducing to the following:

$$H = \frac{DHf}{x_1 y_1}.$$

x_1 and y_1 are distances scaled off the diagram of the principal plane (fig. 22) as shown.

Drill for Semi-graphic Height Determination

BASIC DATA. Requirements are that the height of the aircraft be known, that either the visible horizon appear on the photograph, or that it or the photo plumb point can be determined, and that f be known. The necessary modifications to the construction when working from the plumb point are obvious from the figures. Referring to figs. 21 and 22, drill follows:

DRILL:

(i) On the photograph draw the principal vertical through p at right angles to the visible horizon and meeting it in h'. Scale accurately the distance D (fig. 21), using a millimetre scale and magnifying glass. Mark m, the middle of the height of the feature.

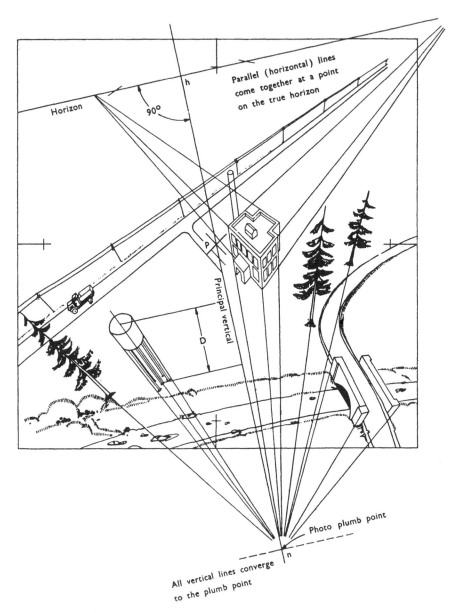

Fig. 21. Drawing of photograph, showing perspective principles.

(ii) On a piece of drawing paper draw Sp (fig. 22) equal to f, and draw the photo-plane hpn at right angles to Sp. With dividers, transfer the distance ph' from the photograph. Join Sh', giving the visible horizon in elevation. (The true horizon lies at h, above this, the angle hSh', in minutes, being the square root of the height of the aircraft in feet. Thus if H is 900 ft., hSh' is

30′. For most purposes it will be sufficient to work from the visible horizon. But for the purpose of instruction, the true horizon should be used.

(iii) Set one point of a pair of dividers at m, midway between the top and bottom of the feature, and take off the distance to the *visible* horizon (there is no need to draw the true horizon on the picture). Transfer m to the photo-plane. From m thus located on fig. 22 drop a perpendicular to the *true* horizon in m_1. Scale x_1 and y_1.

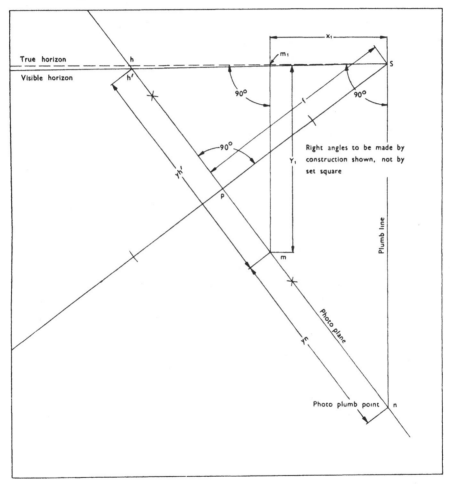

Fig. 22. Elevation of principal plane. Semi-graphical heighting from a single oblique.

(iv) We may now calculate h directly, using a slide rule:

$$h = \frac{DHf}{x_1 y_1} = \frac{4 \cdot 2 \times 900 \times 152 \cdot 4}{102 \cdot 2 \times 100 \cdot 2}$$

$= 56 \cdot 2$ ft. (the last figure is not significant).

Note that D, f, x_1 and y_1 must *all* be in the same units, all in millimetres if we are working in millimetres, or all in inches if we are working in inches. h will come out in the same units as H, feet in this case.

PROCEDURE WHEN THE HORIZON IS NOT VISIBLE. Where the horizon is off the format, or is obscured, it may be located by the Crone construction or by the four-point method. The Crone construction also gives height of aircraft above the feature. These and certain other methods about to be described may be used with a vertical photograph, instead of a base map or ground control. Height of aircraft may also be found readily.

On large-scale low-altitude non-horizon obliques an approximation to the horizon or to the plumb point (the one serves as well as the other) can often be obtained by remembering that parallel lines converge to the horizon, and vertical lines converge to the plumb point. Fig. 21 should be studied carefully. Generally the difference between true and visible horizon may be neglected.

Heighting Grids

INTRODUCTION. It is possible to construct a grid which may be placed over the photograph and from which the heights of vertical objects may be obtained provided only the height of aircraft and depression angle are known. The use of a grid for heighting was suggested by C. H. Seely of the Dominion (of Canada) Forest Service. The following vertical-object technique was worked out in collaboration with the School of Military Engineering, R.E. This grid, and its properties, will more readily be understood after drawing one according to the following drill.

DRILL FOR HEIGHTING GRID CONSTRUCTION. The actual grids are usually drawn up about twice full size, photographed down on glass, and printed on topo-base film. Really careful workmanship is required to ensure a high order of accuracy. To construct the grid for H 100 ft., f 8 in., d 12°, procedure is as follows (fig. 23):

(i) Calculate ph and pn:

$$ph = f \tan d = 8 \times \tan 12° = 1\cdot70 \text{ in.,}$$
$$pn = f \cot d = 8 \times \cot 12° = 37\cdot1 \text{ in.}$$

Mark these data and f, d and H in the upper right-hand corner of the sheet.

(ii) Draw ph produced, as shown, and mark p and h.

(iii) At p erect the perpendicular pS using the geometric construction; mark S so that pS equals f, and join Sh.

(iv) The figure so drawn is the elevation of the principal plane of the projection, and the angle hSp is equal to d. Check this with a good protractor. If a precise check is not obtained, find out why.

(v) Mark the top and bottom of the format, and calculate D, the image height of a 10 ft. object, from the following formula:

$$D = \frac{h}{Hf}\left(fa - a^2 \frac{\sin 2d}{2}\right)$$
$$= \tfrac{1}{80}(8a - 0\cdot2033a^2),$$

where a is the distance, measured on the photograph, from the centre of the image of the 10 ft. object to the horizon, and h is the height of the object, 10 ft. in this case. Calculate D for not less than three positions, i.e. for $a = 1$, 3 and 4 in., covering the format

$$D = \tfrac{1}{80}(8 - 0 \cdot 203) = 0 \cdot 097 \quad \text{for } a = 1 \text{ in.,}$$

$$D = \tfrac{1}{80}(16 - 0 \cdot 813) = 0 \cdot 190 \quad \text{for } a = 2 \text{ in.,}$$

$$D = \tfrac{1}{80}(32 - 3 \cdot 25) = 0 \cdot 359 \quad \text{for } a = 4 \text{ in.}$$

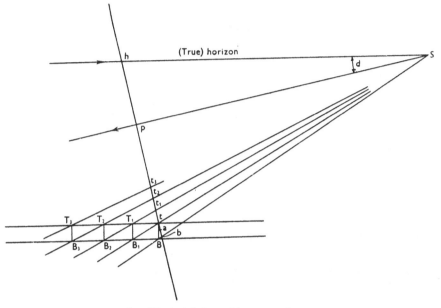

Fig. 23. Heighting grid construction.

The above formula follows directly from that given in the semi-graphic construction. Slide-rule solution to three significant figures is sufficient.

Mark the a positions on the photo-plane.

(vi) At the position of the lowest a, mark $tb = 0 \cdot 359$ in., so that it lies centrally between t and b. This tb must be as nearly precise as possible. Having marked tb, scale it in millimetres using a Leitz magnifier if available, and convert to inches as a check. Example: tb scales $9 \cdot 04$ mm. $= 0 \cdot 356$ in., which is permissible.

(vii) Join St, Sb as shown, and from t drop tB perpendicular to the horizon to intersect Sb produced in B. Now tB may be thought of as a vertical object the photographic image of which is tb. Through t and B draw parallels to the horizon, producing them well to the left. These parallels must be drawn with precision.

(viii) From B_1, the intersection of St produced with the lower horizontal, erect the perpendicular $B_1 T_1$ and join $S T_1$ cutting the photo-plane in t_1.

(ix) Repeat (viii) from B_2, B_3, etc., obtaining the successive image positions t_1, t_2, etc.

(x) As you pass the a position previously calculated, check the photo-plane dimension carefully, as in (vi). The check being satisfactory (it should check to 0·01 in.) continue till the next a, checking again finally at the end of the grid.

Note. When drawing these grids for reproduction and actual use, the a position is scaled and the interval calculated at each third or fourth tb. See also (xiii).

(xi) With the smaller depression angles, the interval converges rapidly as the horizon is approached. Carry the grid up to the point where 10 ft. equals about 2 mm. at the full-size scale—4 mm. on a 2 × enlargement.

(xii) As T and B move out to the left during the construction, and SB becomes too large for convenience, a fresh set of parallels may be drawn through the last located t's, and the construction continued from these new parallels.

(xiii) If the final t to be drawn does not lie near a previously calculated a, check by means of the formula.

(xiv) Now parallels, at right angles to hp, may be drawn to cover the required format and inked in. The 10 ft. intervals are rather large (maximum $\frac{4}{10}$ in.) for interpolation of heights on the photo; they are therefore sub-divided into five equal parts of 2 ft. This subdivision should be continued until the 10 ft. interval reduces to about 5 mm.

For the same depression angle, the grid for other focal lengths may be obtained by photo-reduction of the original to a different scale—thus only one original is required for each d.

Fig. 24 shows two heighting grids for the 8 in. cone, one a horizon oblique d 10°, the other a non-horizon oblique, d 45°. Note how, on the 10° oblique, the interval representing 10 ft. converges rapidly—reaching zero at the horizon—while the 'vertical scale' of the 45° is nearly constant across the whole format. Reproduction is about half-size of the grid, which is a trans-parency on topo-base film.

NUMBER OF GRIDS REQUIRED. The size of the 10 ft. interval is a quadratic circular function of d, and accordingly its rate of change with respect to d is rather complex. For our purposes, grids calculated for such intervals of d as will result in 10% increments are satisfactory. The average *photogrammetric* error will then be $2\frac{1}{2}$% or, for example, 3 inches in the deter-mination of a 10 ft. height. This is closer than, ordinarily, is justified by the basic data. At p the interval is 0·40 in. for d 45° and f 8 in. Accordingly, the next d should be such as to give an interval, at p, of 0·40 − 0·40/10 = 0·36, the third 0·36 − 0·36/10 = 0·324, and so on. This results in the grids for the following values of d in degrees: 45, 33, $27\frac{1}{2}$, $23\frac{1}{2}$, $20\frac{1}{2}$, 18, 16, $14\frac{1}{2}$, 13, $11\frac{1}{2}$ and 10. The grid for a particular d will also apply, if reversed, to its com-plement and can be used where d is in excess of 45°. It should be pointed out that the 10% increments are at p, above p the increment is greater,

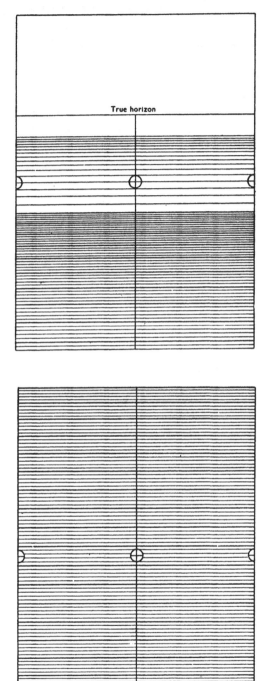

True horizon

Fig. 24. Heighting grids, one-half actual size. The grids are printed on film, and laid over the photographs.

below p it is smaller. It is stressed that this interval is practical and is consistent with the strength of the basic data. A family of eleven grids, one for each of the above depression angles, is provided for 5, 6, 8, 12 and 20 in. lenses. All are drawn for a standard height of 100 ft. above the ground.

The grids are ruled with horizontal lines at a distance apart representing the image height of a 10 ft. vertical object photographed from 100 ft. above the ground, the 10 ft. lines being further divided into 2 ft. rulings where the scale is open enough to permit this. The height factor is the true height of aircraft (above the object) divided by 100. The procedure, briefly, is to lay the grid on the photograph, take the reading, and multiply by the height factor. The result is the required height of the object. The most complex mathematics is the multiplication of the grid reading by the height factor.

HEIGHTING GRID DRILL, VERTICAL OBJECTS. These grids were first constructed in 1944 for use in the North-West European campaign. 1:25,000 (or larger) photogrammetric maps with 5 or 10 m. contours had been compiled previous to the invasion, and the use of heighting grids in conjunction with these base maps is contemplated in the following drill:

(i) From the 1:25,000 sheet estimate the elevation at the object you are heighting—the map gives metres above sea-level, and this must be changed to feet. This you may do by changing metres to yards using the map scale (which shows metres and yards) and multiplying by three, or if you prefer you may use the following conversion table.

TABLE 5. *Conversion table, metres to feet*

Metres	Feet	Metres	Feet
1	3	10	33
2	7	20	66
3	10	30	98
4	13	40	131
5	16	50	164
6	20	60	197
7	23	70	230
8	26	80	263
9	30	90	295
10	33	100	328

Example. Convert 158 m. to feet.

Metres	Feet
100	328
50	164
8	26

158 m. = 518 ft.

Round this to the nearest 10 ft., i.e. 520.

(ii) The height of aircraft above sea-level is shown on the heighting photographs. Subtract from this the elevation at the object and divide by 100 to obtain the height factor. Thus

H from annotation   900 ft.

Elevation at object   520 ft.

Hence

Height of aircraft   380 ft. above the ground, and height factor is 3·8.

(iii) If the principal point is not shown on the print, mark it with a cross in pencil by joining the fiducial marks or, if there are none, the corners of the print. Draw the principal vertical. This direction may be judged from

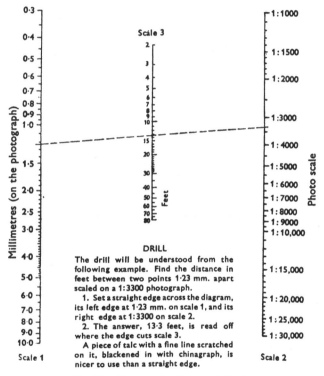

Fig. 25. Photo-scale alinement chart.

trees, buildings, etc., near p. A study of fig. 21 should make it clear how to draw this line. The principal vertical is at right angles to the horizon if the horizon is on the picture—which will not usually be the case. Fig. 21 is drawn for a very wide-angle lens in order that the perspective shall be exaggerated.

(iv) Using a magnifying glass and a pricker, prick the top and the bottom of the vertical distance you wish to find; for example, the height of the water tower of fig. 21. A fine pin-prick is all that is needed; just enough to let a spot of light through. You must take the middle of the top of the tower, not an edge. Imagine a plumb line hanging from the top mark you have made, it should hit the ground at the bottom mark. The grid measures the distance D (fig. 21), and converts it to the height of the object in feet.

(v) The photograph is marked with the depression angle d, and the focal length f. On the print lay the grid (of correct f) whose depression angle is nearest to that shown. Aline it on the principal vertical with p of the grid (the circle) on p of the photograph.

Note. If the depression marked on the photograph is in excess of 45°, subtract the reading from 90° and use the grid nearest this, placing it upside down on the print.

Example. d 57°, 90 − 57 = 33. Use the 33° grid, but upside down.

(vi) Keeping the grid in position, hold the print up to the light, and read the grid height directly in feet. To facilitate this, you may move the grid slightly so that one spot of light is cut centrally by a grid line.

(vii) Multiply this reading by the height factor and the answer is the required height. This multiplication may be done on the alinement chart (fig. 26).

Example. Reading 7·8 ft., height factor, from (ii) 3·8, required height 7·8 × 3·8 = 30 ft., to the nearest foot since the decimal scarcely is significant.

(viii) Always use the common-sense checks. Measure anything (a house, a telegraph pole) the height of which you know, or can estimate. For example, a pole measured 6·2 × 3·8 = 24 ft., and a hut 4·0 × 3·8 = 15 ft., on the photograph above. This is reasonable. If the height does not come out as it should you may proceed as follows—but before doing so be sure you have not made a blunder, such as using the wrong depression angle, wrong focal length, or wrong height above ground. Assumed height of pole 25 ft., reading on grid 5 ft., hence height factor = $\frac{25}{5}$ = 5. That is to say, work out a new height factor by dividing the known (or assumed) height of a test object by its grid height.

GROUND TEST, HEIGHTS OF VERTICAL OBJECTS. Table 6 gives the result of a ground check on photogrammetric heights determined both by the semi-graphic method and by means of the heighting grids. These are vertical objects.

Since d was small—and the horizon obscured by detail—both the horizon and height of aircraft were determined by the Crone construction. The photogrammetry was by the Service Topographique de l'Armée Belge, and the several figures of columns 4 and 5 of the table are results obtained by different individuals in measurement of the same height. Ground check was by the Field Survey Company, R.C.E.

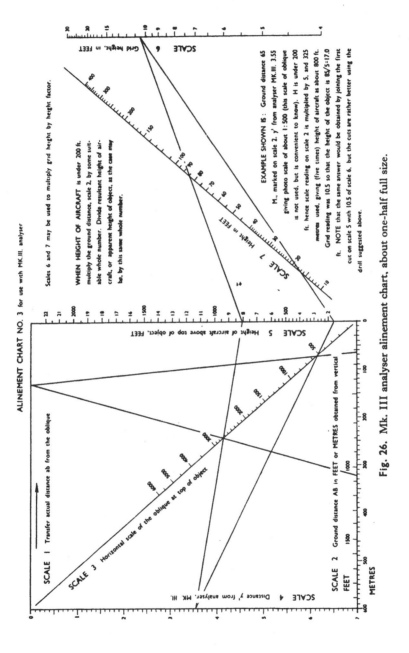

Fig. 26. Mk. III analyser alinement chart, about one-half full size.

TABLE 6. *S'Hertogenbosch Tests. Vertical Objects*

Ground check on oblique heighting by the Service Topographique de l'Armée Belge, s'Hertogenbosch, Holland. Date of sortie, 26 September 1944

Height of aircraft (ft.)	d, measured by the Crone method	Feature	Heights in ft.			Error (ft.)
			Semi-graphic	Grid	True	
875	14° 01′	Building	25·7	26·2		Semi-graphic +1·2
			26·2	26·7		Grid +1·3
			26·7	26·4		
			26·6	26·2		
		Mean	26·3	26·4	25·1	
875	14° 01′	Building	28·4	28·9		Semi-graphic −0·1
			28·4	28·8		Grid −0·1
			28·6	29·2		
			28·5	27·0		
		Mean	28·5	28·5	28·6	
820	14° 00′	Building	60·5	62·0		Semi-graphic +0·2
			63·4	64·3		Grid +1·0
			62·3	62·6		
			62·0	64·0		
			63·3	62·8		
		Mean	62·3	63·1	62·1	
865	12° 56′	Building	35·9	34·6		Semi-graphic +2·8
			36·6	35·4		Grid +2·2
			35·9	35·0		
			34·7	35·3		
			36·6	36·3		
		Mean	35·9	35·3	33·1	
865	12° 56′	Building	20·1	19·1		Semi-graphic +2·2
			22·9	20·7		Grid +1·8
			20·6	19·9		
			20·1	21·0		
			20·5	21·2		
		Mean	20·8	20·4	18·6	
865	12° 18′	Building	51·0	51·0		Semi-graphic +3·2
			51·0	50·6		Grid +3·2
			51·3	51·5		
			51·0	51·0		
			51·3	51·2		
		Mean	51·1	51·1	47·9	
865	12° 18′	Building	45·2	45·4		Semi-graphic +3·5
			45·3	45·4		Grid +3·5
			45·1	45·0		
			45·2	45·0		
			45·3	45·0		
		Mean	45·2	45·2	42·7	
1645	15° 18′	Building	62·0	61·7		Semi-graphic +1·3
			61·8	61·7		Grid +1·3
			61·9	63·3		
			65·7	65·0		
			65·7	65·8		
		Mean	63·4	63·4	62·1	
1645	15° 18′	?	28·5	29·6		
			28·3	29·6		
			28·8	29·2		
			28·3	28·0		
			28·4	28·8		
		Mean	28·4	29·0	Destroyed	
865	12° 18′	Clearance, road to railway arch invert	—	14–15	13·6	About a foot, by grid
Arithmetic mean error, all observations				1·9		

Determination of Height of Aircraft, and Measurement of Sloping Objects, by means of the Canadian Oblique Analyser, Mk. III

BASIC DATA.   So far we have treated only the heighting of vertical objects, for which the basic data of depression angle and height of aircraft above feature were given. Where these are known, or can be determined, the orders of accuracy of table 5 can be obtained rapidly and, substantially, without further calculation. If the depression of the camera axis be set at some arbitrary, known, value between say 30 and 40°, then the ordinary tilts either of civil or of wartime flying can be neglected, and d taken as given. At $d = 45°$ plus or minus $12\frac{1}{2}°$ tilt has the same effect upon heighting grid precision as has a tilt of plus or minus $1\frac{1}{2}°$ at $d = 10°$. The importance, therefore, of a large depression angle is apparent where heighting is to be by this method.

Given a large-scale, reasonably high altitude, vertical we may use this as control and obtain the height of aircraft, and as the position of the true horizon, by means of the Crone construction. The scale of the vertical is taken as f/H-h, H being accepted as the altimeter reading, and h being guessed as well as possible. Note that with photography from 15,000 ft. an error of 300 ft. in the estimation of H-h will cause only a 2 % error in measured height, less than 5 inches in a 20 ft. object. With photography at 30,000 ft. a 300 ft. error in H-h introduces only a 1 % heighting error. Thus we can use the method in entirely unmapped areas, either in peace or in war, since usually we can tolerate larger errors than these. However, the Crone construction is rather lengthy, and some more rapid method of obtaining height of aircraft would be of advantage. The newly developed narrow beam radar altimeter gives H-h directly. This instrument is known as the APR.

Where the heights and gradients of sloping objects, such as banks, dykes, and the like, are required, we measure the *apparent* vertical height of the feature using the means just described. Then we obtain the horizontal, or plan, dimension from a vertical photograph of the site. From these data the oblique analyser will give a cross-section scale 1 in. to 10 ft. without calculation. The analyser will also give height of aircraft in feet quite readily, and this height, even if determined from a horizon a couple of centimetres in error, will not cause a corresponding error in the height or slope of the feature—since the height so obtained is in sympathy with the incorrect horizon.

USE OF THE MK. III ANALYSER.   Given an oblique photograph, either horizon or non-horizon, and a vertical photograph of the site, H-h of the vertical, hence its scale, being known within the limits discussed above, the drill follows.

The depression angle of the oblique may be taken as the depression angle to which the camera axis has been set in the aircraft. H-h of the oblique photography is not known within tolerable limits. The method is applicable to ranges of H-h from zero to several thousand feet, to all depression angles, lenses and formats—a photo-specification is, however, given on p. 52, and optimum results will be obtained from such photography.

Drill, Stage I. Transference of principal vertical, scale and section lines:

(i) The object being sectioned, say a river bank, should run more or less in the cross-direction, not up and down the format. Through the feature, and radial from h, draw a fine line. The section given by the analyser will be along this line, which ordinarily will not be the line of steepest slope. Through the top of the feature draw a line parallel to the horizon; this line is the scale line.

(ii) Transfer the principal vertical and the section line to the vertical photograph; note that on the vertical these lines are parallel, since we drew the section line radially from h.

(iii) On the vertical choose two well-defined image points A and B such that the line AB is at right angles to the transferred principal vertical, or nearly so. Transfer these points to the oblique and mark them a and b. Join ha and hb produced to cut the section line in a′b′. Measure AB and a′b′ either in inches or in millimetres as is convenient—so long as both are measured in the same units. Then horizontal scale of the oblique at the top of the feature is   Scale of vertical × a′b′/AB.

The Mk. III analyser alinement chart (fig. 26) should be used for this calculation, which is further facilitated (if fig. 26 is used) by the photo-scale alinement chart (fig. 25). Actually we do not need to know the numerical value of the scale of the oblique, but it is a convenience, even if only to check whether the scale is to specification. The diagonal of the left-hand side of the alinement chart (fig. 26) actually need not be graduated.

Drill, Stage II. Use of the analyser (fig. 27) to determine height of aircraft above the feature.

(i) Transfer the distance ph of the oblique to p′h′ of the analyser, p′ being vertically below the intersection of the relevant focal length segment with the H scale. Join p′ to o, to intersect the f segment in p and the degree segment in d, giving the depression angle if required. Where d is given, ph need not be transferred.

(ii) On the oblique set dividers to the y co-ordinate of the top of the feature, and mark t on the analyser so that pt = y. Join t to o. From t draw tt′ perpendicular to the H scale, or simply read off the distance of t below this scale. Now, tt′ represents the height of aircraft above the feature, its scale being that of the oblique at the scale line. Transfer tt′ (= y′) to scale 4 of fig. 26, join with the scale 3 cut giving oblique scale, and produce to cut scale 5 where height of aircraft may be read directly.

If the horizon be incorrect by a centimetre or so, the height just determined is in sympathy with the incorrect position—and so will introduce no sensible error.

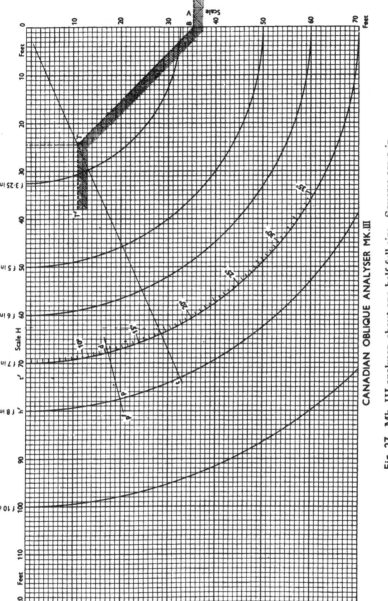

CANADIAN OBLIQUE ANALYSER MK. III

Fig. 27. Mk. III analyser, about one-half full size. Squares are 1 in.

Drill, Stage III.   Sloping objects.

The foregoing stages I and II are applicable as well to vertical objects. The procedure would be as before, using a height factor from the H just found and the grid of the depression angle indicated. The method is thus applicable to entirely unsurveyed country. For sloping objects, drill is as follows:

(i) Prick, very carefully, the top and bottom of the feature on the section line, and obtain this distance, in apparent feet, by means of the heighting grid. Multiply the grid reading by the height factor—or use scales 6 and 7 of fig. 26 for this purpose.

(ii) On the analyser, set off this apparent height in feet at B on scale V.

(iii) On the vertical determine by careful measurement the horizontal dimension of the feature on the section line. Use a Leitz glass and measure to the nearest half-tenth of a mm. Knowing the scale of the vertical, convert this to feet by means of the photo-scale alinement chart (fig. 25).

(iv) On the analyser, on scale H, set off this distance in feet and drop a perpendicular to intersect t_0 in T. Now T'TB is a true section of the feature along the section line; scale 1 in. equals 10 ft. True height may be read from the V scale, being the vertical distance between T and B.

Note that this section is along the section line marked in I (ii) above. The means of converting this section to a right section are obvious.

SPECIFICATION FOR HEIGHTING PHOTOGRAPHY.   The accurate identification of the top and bottom of a feature such as a dyke is not easy—nor is it easy to transfer these points from the oblique to the vertical. In order that this may be done with precision it is *essential* that both obliques and verticals be stereoscopic pairs at as large a scale as conditions will permit.

There are the following variables: angle of cone, size of format, height of aircraft, exposure interval, speed of aircraft, depression angle, exposure considered in relation to speed and image movement—and exposure considered from the photographic point of view. If good results are to be obtained, these factors must be calculated. The calculations are capable of simple alinement chart solution, so that the task is not so formidable as might be expected.

The following general essentials are stressed:

Oblique and vertical photography should be simultaneous to avoid error from tidal or other changes in water-levels.

Obliques to be lateral.

Vertical photography must be large scale, preferably of the same order as that of the oblique at the site being heighted, the scale to be obtained from high-altitude long focal length photography. It is of course stereoscopic.

The lateral oblique photography should be stereoscopic, minimum 60% overlap at p is specified. f, interval, depression angle, and speed of aircraft must be specified to attain this.

Large depression angles are preferred.

Scale of oblique at p should be 1:5000 or larger—the object being in the foreground.

The lens angle is relatively unimportant as such—there is no particular case for wide angle in this application. Very good results can be obtained with a cone as narrow as 20 in. on 5 × 5 format—although it is not necessary to go to this extreme.

A focal-plane shutter is quite satisfactory provided that the screen travel across the format—not in the y-axis. The direction of travel should be such as to reduce image movement, i.e. in direction of travel of the aircraft.

GROUND TESTS ON HEIGHTING SLOPING OBJECTS. During February and March 1945 values of heights and of slopes of dykes and of river banks on the Rhine between Emmerich and Xanten were required by Engineer Intelligence of the First Canadian and Second British Armies. This information was obtained from obliques using heighting grids and the analyser. The Crone construction was used for horizon and height of aircraft, the Mk. III analyser not having been devised at that time. Photography was not according to the specification above; the results therefore should be considered as indicative of what can be done under adverse conditions. Ground measurements were made in April 1945, after the crossing, and tables 7 and 8 give photogrammetric and measured heights.

TABLE 7. *Stereoscopic Oblique Cover, Sloping Objects*

Heights more than 10 ft.			Heights less than 10 ft.		
Photo	True	Error	Photo	True	Error
12	12·8	0·8	9	8·2	0·8
12	14·1	2·1	10	7·6	2·4
12	12·1	0·1	9	7·6	1·4
12	11·0	1·0	5	3·5	1·5
13	12·2	0·8	7	5·3	1·7
10	11·1	1·1	Average error 1·6 ft.		
14	14·5	0·5			
13	12·2	0·8			
19	17·0	2·0			
12	12·5	0·5			
16	14·0	2·0			
13	12·0	1·0			
18	15·6	2·4			
20	18·6	1·4			
13	11·6	1·4			
12	11·6	0·4			
15	17·5	2·5			
16	13·8	2·2			
13	13·8	0·8			
17	17·5	0·5			
16	13·7	2·3			
Average error 1·3 ft.					

One should bear in mind that, at 1:5000, 10 ft. is but 0·6 mm., and that therefore a small scaling error either on the oblique or on the vertical will introduce a large error in the height of a sloping object. All photographic distances must therefore be measured with extreme care.

Fig. 28 is a reproduction of one of the obliques used to obtain the above measurements. For operational reasons flying height is very low, often

TABLE 8. *Non-Stereoscopic Oblique Cover, Sloping Objects*

Heights more than 10 ft.			Heights less than 10 ft.		
Photo	True	Error	Photo	True	Error
4	—	—	12	8·6	3·4
11	15·0	4·0	6	9·8	3·8
14	18·0	4·0	13	10·0	3·0
9	11·4	2·4	9	8·7	0·3
29	3·0	2·0	9	7·4	1·6
24	27·0	3·0	8	7·8	0·2
17	13·6	3·4	12	8·0	4·0
20	24·3	4·3	10	7·8	2·2
18	15·0	3·0	7	10·3	3·3
10	10·4	0·4	4	4·4	0·4
14	15·1	1·1	7	7·2	0·2
25	25·1	0·1	7	6·6	0·4
11	10·8	0·2	9	8·4	0·6
18	18·1	0·1	10	6·5	3·5
11	11·5	0·5	11	8·9	2·1
17	13·2	3·8	9	8·8	0·2
15	16·5	1·5	10	8·5	1·5
19	16·3	2·7	Average error 1·8 ft.		
20	17·2	2·8			
Average error 2·2 ft.					

under 100 ft.—consequently the depression angle must be small. Speed of aircraft is very high. Fig. 29 shows the sections at the sites designated in fig. 28.

Summary of the Uses of Obliques

HORIZON OBLIQUES. Horizon obliques of flat terrain provide a rapid method of mapping at small or medium scales; they are well suited to shore-line mapping. High obliques of rough mountainous country may be used to provide vertical and horizontal control. This control may be established rapidly, and extended to considerable distances. High obliques may be used to map such mountainous country, but as a large number of points must be cut in the procedure is slow, and dead ground may leave important areas blank.

INTERPRETATION FROM HORIZON OBLIQUES. Interpretation is seemingly easy, as obliques present a familiar picture of the ground. Stereoscopic pairs are virtually essential and should be insisted upon. Dead ground is a difficulty—both minor and major features of importance, which are apparent even on a single vertical may, on a high oblique, be obscured completely by topography or vegetation.

NEAR VERTICALS AND LOW OBLIQUES. With h/H small, near verticals are quite suitable for mapping planimetry. With sketchy vertical control

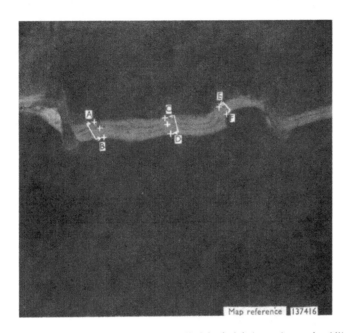

Map reference ▮137416

Fig. 28. Low altitude non-horizon oblique supplied for heighting prior to the Allied crossing
of the Rhine. (R.A.F. photograph.)

To face p. 5

good form-lining can be done, but precise topography is not obtainable by simple means. With h/H large there is no simple non-instrumental method of using low obliques or near verticals for mapping purposes.

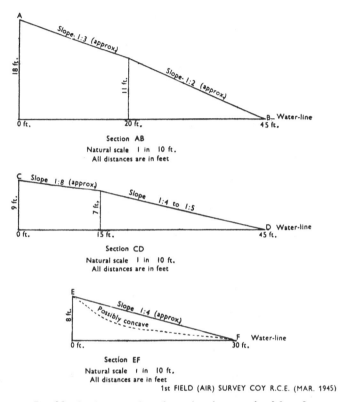

Fig. 29. Analyser sections from the photograph of fig. 28.

Interpretation from near verticals differs little, in practice, from that of verticals.

There is a considerable forestry application both of horizon and of non-horizon obliques in the measurement of tree heights, and an important military application in the measurement of heights and slopes.

CHAPTER 5

VERTICALS: FUNDAMENTAL CONSIDERATIONS OF COVER, PARALLAX AND STEREOSCOPY

Verticals

INTRODUCTION. Certain of the methods described in preceding sections are often spoken of as being techniques of a single photograph. This is perhaps a misnomer (see the definition of p. 1), since, in all cases involving mapping of an area larger than the extent of a single exposure, overlapping pairs are used. Both with non-verticals and with verticals interpretation should be done stereoscopically if advantage is to be taken of the full value of the photography.

Techniques using vertical photography are, similarly, often described as mapping from stereoscopic pairs. This term, too, is misleading. For while we measure parallax differences with the aid of a stereoscope, and sketch contours stereoscopically, still, all the simple mapping methods described in this and succeeding chapters accept shape, both of detail and of contour (sensibly) as shown upon a single overlap of the pair. Not until 'automatic' plotting machinery is employed do we actually map from stereoscopic pairs, obtaining the true orthogonal projection of the relief model.

OVERLAP. When it is required that an area be covered by vertical photographs, the aircraft flies as nearly as possible in a straight course at the constant height indicated by scale and other considerations. A strip of photographs is exposed in such a manner that each principal point appears on the preceding and on the following picture, and each thus contains three principal points, its own and that of each adjacent picture of the strip. The foregoing results in the necessity of a minimum fore-and-aft overlap of 50%. To allow for tilt and relief, for variations in height of aircraft above the ground, and for crab (q.v.), it is usual to specify 60% fore-and-aft overlap,* and the exposure interval is adjusted accordingly.

Fig. 30 should make the arrangement clear. Note how the overlap decreases with the high ground at P_1, and increases in the valley at P_2. The whole area to be mapped is covered by a parallel series of such strips, and it is customary to specify that there be a lateral overlap* of 20–35 %.

CRAB. The line of flight of an aircraft is the resultant of the air-speed vector and the wind-velocity vector—like a rowboat crossing a stream the centre-line of an aircraft in flight does not point in the direction of motion

---

* The American Society of Photogrammetry recommends the use of the terms 'forward lap' and 'side lap' in lieu of the above designations.

relative to the ground. The angle between the motion relative to the air and the motion relative to the ground is the angle of crab (fig. 31). To overcome this the camera is rotated in azimuth through an angle equal to the angle of crab; thus the line of flight on the photograph becomes parallel to the format.

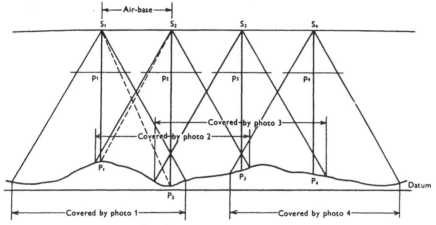

Fig. 30. Vertical cover, overlap.

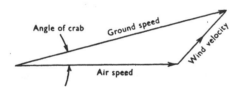

Fig. 31. Velocity vectors, and angle of crab.

AIR-BASE. Referring to fig. 30, S_1S_2—the distance between successive exposures—is called the air-base. This is equal to the distance between P_1 and P_2. On photo 1 the distance p_1p_2 is the photo air-base, b. p_1p_2 on 1 is not the same as p_2p_1 on photo 2 unless the ground points P_1 and P_2 are at the same elevation.

On the photograph, the line joining the image positions of successive p's generally is called the base-line. The base-line is sometimes referred to as the direction, or azimuth, line. Base-lining a photograph consists in establishing the direction line from its principal point to the images, on the same photograph, of the principal points of the adjacent overlapping photographs. The first and last photograph of a flight will have one such direction line, intermediate photos each having two.

DRILL FOR BASE-LINING. Base-lining drill (fig. 32). The principal point (if not marked) is obtained by joining opposite fiducial marks. Use a needle and scribe a fine cross as shown. If the principal point of photo 2 falls exactly upon a point of well-defined detail, then its position can be identified readily on photo 1, and the base-line p_1p_2 drawn without further ado. As this seldom happens, we proceed as follows:

(i) Identify p_2 approximately on photo 1, and aline a straight edge between this point and p_1.

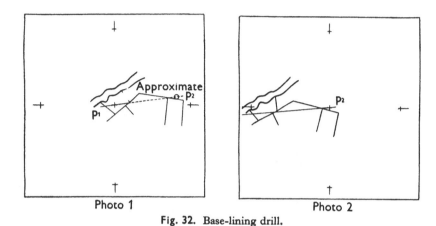

Fig. 32. Base-lining drill.

(ii) Scribe (use a fine needle) a short line through p_1 as shown. A small error in identifying p_2 on 1 will not affect this part of the line.

(iii) Now, instead of trying to identify p_1 on photo 2, it is normally possible to select some minute piece of detail which lies exactly on the short scribed line. Transfer this point to photo 2, and join to p_2, giving the required base-line.

(iv) The base-line on photo 1 is now fixed by the same means, using the right-hand end of the base just drawn on photo 2.

(v) The base-lines as drawn should cut identical detail on each photograph.

(vi) The work must be carried out very carefully. A precise check is obtainable by stereoscopic means described on pp. 81 and 94.

Parallax

A PHYSICAL EXAMPLE OF PARALLAX. Hold your finger out at arm's length and close the left eye. Look at your finger and, at the same time, at the wall of the room. Now open the left eye and close the right. You will observe that your finger seems to have moved with relation to the wall, whereas, in fact, it has remained stationary. The explanation is that this apparent motion results from the fact that the left eye and the right eye are about $2\frac{1}{2}$ in. apart. Another thing to notice is that this apparent motion is parallel to the eye-base, the eye-base being defined as the imaginary line joining the perspective centres of the two eyes. This may be illustrated by tilting the head and repeating the above experiment.

PARALLAX ON VERTICALS. Now examine fig. 33. S_1 and S_2 represent successive positions of the perspective centre as exactly vertical photographs are taken. The altitude of the aircraft is supposed to be constant, that is, the air-base S_1S_2 is horizontal. The ground point A photographs at a_1 on photo I and at a_2 on photo 2.

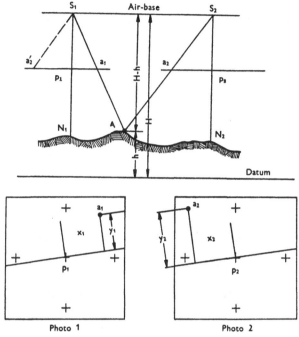

Fig. 33. Parallax.

The photo base-lines are shown, the x ordinate of the point a on photo I is x_1 and on photo 2 is x_2. What is the apparent motion of the image point a, measured in a direction parallel to the base-line, on these two successive pictures? Obviously $x_1 - x_2$. (*Note.* In the figure x_2 is negative, since it is to the left of the origin p_2; the usual sign convention is followed.) Now for our purposes we define parallax as follows: The parallax of a point is the change in position of its image on two successive exposures of an overlap. This change is measured parallel to the base-line. Remembering the apparent motion of your finger on the wall this conception of parallax is readily grasped.

WANT OF CORRESPONDENCE. Place one point of a pair of dividers on a_1 on picture I and stretch the base-line. The dividers are now set to y_1 (fig. 33). Set on a_2, and if the y ordinates on the two pictures do not correspond, then there is 'want of correspondence'. The amount of this want of correspondence, denoted by K, is $y_1 - y_2$. K is often called y parallax, particularly by American writers.

Assuming, as we have, that both photographs are truly vertical, and that

S_1 and S_2 are at the same height above datum, this condition cannot occur (except from distortion.) Under these conditions, the y ordinates on the two photographs must be equal, must correspond. Want of correspondence may arise from any of the following circumstances, either singly or in combination:

Tilt of photo 1, either fore and aft or lateral.

Tilt of photo 2, either fore and aft or lateral.

Want of parallelism of the air-base, that is, S_1 and S_2 not being at same height above datum. Distortion of any kind.

PARALLAX EQUATION. On fig. 33 through S_1 draw S_1a_2' parallel to S_2a_2, then $p_1a_2'=p_2a_2$; hence $a_2'a_1$ is, according to our previous definition, the parallax of A. This is denoted by P.

The triangles S_1S_2A and the small triangle $a_2'a_1S_1$ are similar. The altitude of the smaller triangle is the focal length f, and that of the larger one (H-h). Denoting the air-base by B, we have $P/f=B/(H-h)$. Hence

$$P=fB/(H-h).$$

This is known as the parallax equation.

CALCULATION OF PARALLAX. We can now use the parallax equation to calculate the difference in position between the image of a point as it appears on two successive overlapping photographs, and we can check by measuring the parallax and comparing this with the calculated value.

Exercise 7. You are given a vertical pair and a 25,000 map * of the area; f is known. It is required to calculate the parallax of a point by means of the parallax equation, and to compare this result with the measured parallax. A slide-rule may be used.

(i) Pick two points at the same (or nearly the same) elevation on the map and identify them on the photograph. Both points must be on the same photograph. Measure the distance between these points on the map and on the photo and so calculate the photo scale. Several pairs of points should be chosen, one or two pairs on each photograph. The distances between the points may be measured in any units, so long as the same units are used for measuring on the map and on the photograph (measurements in mm.):

On map scale 25,000	On photo	Scale denominator
34·4	34·6	34·4/34·6 × 25,000 = 24,850
62·0	63·5	= 24,760
43·6	44·0	= 24,790
		Mean 24,800

Now calculate the height of the aircraft above datum, using the formula scale = f/12 (H-h). In this example f is 6 in., hence:

* In the absence of a suitable base map, you will be given co-ordinates and elevations of ground control points.

$S = f/12 \, (H-h)$, $(H-h) = f/12S$

$6/(12 \times 24,800) =$ 12,400 ft.

h, elevation (or mean elevation) of scale points =   100 ft.

$H = 12,500$ ft. above datum.

(ii) Base-line the photographs.

(iii) Determine the air-base b by meaning the photo air-bases and multiplying by $(H-h)/f$, h being in this case the mean of the heights of the principal points, which information is obtained from the map:

h of p_1	50 ft.	b_1	52·1 mm.
h of p_2	190 ft.	b_2	51·5 mm.
Mean	120 ft.		51·8 mm.

$B = b \times \dfrac{(H-h)}{f}$, the units of which are feet

$$= \frac{51·8}{25·4 \times 12} \times \frac{(12,500 - 120 \times 12)}{6}$$

$$= 4210 \text{ ft.}$$

Note that both f, which was given in inches, and b, which was measured in millimetres, are changed to feet (25·4 mm. = 1 in.).

(iv) Identify a point on the map and on the photo, one of the points of (i) will do. We now have enough information to calculate its parallax directly from the parallax equation

$B = 4210$ ft. from (iii), $H = 12,500$ ft. from (i), $f = 6$ in., $h = 120$ ft., elevation of point from map.

Parallax in our work is invariably measured in millimetres; therefore in the parallax equation $P = fB/(H-h)$, focal length must also be in millimetres. B and H must be in the same units, that is, both in feet if we are working in feet or both in metres if we are working in metres:

$$P = fB \, (H-h) = \frac{6 \times 25·4 \times 4210}{(12,500 - 120)} = 51·8 \text{ mm.}$$

(v) Now measure the x ordinates of the point on each print. From our definition of parallax, $P = x_1 - x_2$, $x_1 = 12·2$, $x_2 = -39·7$. (Note that x_2 is usually minus.) Hence the measured parallax is $12·2 + 39·7 = 51·9$ mm.

(vi) The calculated parallax and the measured parallax will usually agree within one- or two-tenths of a mm. If they do not agree, there is probably relative tilt, as will be evidenced by the presence of K.

(vii) Measure the y ordinates, and determine K.

CALCULATION OF HEIGHT DIFFERENCES FROM PARALLAX DIFFERENCES. The purpose of the foregoing is to show that parallax is not an abstruse mathematical conception, but that it—and want of correspondence—are simple real things that you can measure with a school centimetre scale.

Exercise 8. On the same pair, let us consider a point at 620 ft. above datum. Such a point is 500 ft. nearer the camera than the first point whose elevation was 120 ft. Will its apparent motion be more or less than that of the first point, and accordingly will its parallax be larger or smaller? Between these two points, what elevation difference is represented by 1 mm. parallax difference?

Applying the formula we have

$$P = fB/(H-h) = 152 \cdot 4 \times 4210/(12,500 - 620)$$
$$= 54 \cdot 0 \text{ mm.}$$

The parallax of the point of higher elevation is thus greater.

The difference in parallax between these two points is $54 \cdot 0 - 51 \cdot 8 = 2 \cdot 2$ mm. Evidently, then, on these photographs and between these two points, $2 \cdot 2$ mm. $= 500$ ft., so that 1 mm. change in parallax corresponds to about 227 ft. change in elevation. This relation, so many feet to 1 mm. difference in parallax, may be considered as parallax scale.

From what has been said, and from these two exercises, it becomes apparent that, by measuring parallax differences between points, we can quite readily calculate height differences. Evidently the measurement of these parallaxes with a centimetre scale is rather crude, but there is an instrument which, in conjunction with a simple stereoscope, will measure differences in parallax nicely to two- or three-hundredths of a mm. This instrument, known as a parallax bar, is described on p. 66.

Binocular Vision and the Stereoscope

PHYSIOLOGICAL CONSIDERATIONS. The physiology of binocular vision is complex, so complex that authorities differ on just how the mechanism works to make us arrive at the mental impression of space, of three dimensions. *Why* is much simpler. It is because we have two eyes spaced some distance apart, and so get an image from a different viewpoint on the retina of each eye. The blending of these different views by some intricate physiological means gives us the ability to see in three dimensions. It is like the way that a ray can be drawn to an unknown point from each of two overlapping obliques, the intersection fixing the point in three dimensions.

PARALLAX AS AN ANGLE. It will be seen that the strength of this three-dimensional impression is dependent upon the angle subtended by the eye-base at the distant object. Strictly speaking this *angle* is the parallax of the object, but we will adhere to our former definition. Though a prolonged discussion of binocular vision is unnecessary here, there are a few points to be considered.

The eye can appreciate angular differences as small as 30″ of arc. The eye-base of an average adult is about 2½ in.—about 60 mm.—it thus subtends 30″ at an object some 440 m. away. Therefore judgement of distances greater than this is not binocular at all; it can be done just as well by a man with only one eye. Much of our space impression comes from experience accumulated, sometimes painfully, from infancy—a young baby is not much good at judging distances; watch him try to grasp an object beyond his reach. Visual angles, the unconscious comparison of known to unknown sizes, and perspective (really the same thing) are not binocular, but greatly aid us in the judgement of distance. At any rate, however it is done, *we can* see in three dimensions with our two eyes.

If you are in an aircraft, say at 5000 ft., the landscape below looks flat and featureless; in fact, it looks just like a single vertical photograph. This is true even though the country may be quite undulating. Why? At 5000 ft. the angle subtended by the eye-base is 8·60″. If there is a mountain 1000 ft. above the surrounding country, the angle subtended at its peak will be 10·75″, or only 2·15″ more. This 2·15″ difference is 14 times too small for the eyes to notice; thus we could not tell the difference between the mountain and the plain, and landscape will therefore seem flat. *Note.* Light and shadow effects, combined with the knowledge that you are over a mountain, might make you think there is relief even though you cannot really perceive it.

WHAT A GIANT WOULD SEE. Now suppose there is a giant a mile tall with eyes a thousand feet apart. What would he see if he had the same kind of eyes as ours? The angles this time are 11° 26′ and 14° 15′ respectively, not the few seconds they were to us, and the difference is 2° 49′. He would then see the relief of the mountain quite clearly; in fact, to him it would look like an anthill 2½ in. high looks to us when viewed from a distance of 12½ in. (see fig. 34).

Exercise 9. Verify this statement.

Tan of ½ angle subtended by 2½ in. eye-base is 1·25/12·50 = 1/10. Angle is 2 × 5° 43′ = 11° 26′.

Tan of ½ angle subtended by eye-base at 10·00 in. (the distance from the eye to the top of the anthill) is 1·25/10 = 0·125, angle is 2 × 7° 07½′ = 14° 15′.

These are the same angles that the giant saw.

Let us photograph this mountain from two positions 1000 ft. apart, as shown at the right of fig. 34. Now the camera, in these two positions, will serve as the giant's eyes if we take these photographs and look at them in a stereoscope. A stereoscope is just an easy way of looking at the pictures so that the left eye sees only the left picture, and the right eye only the right, with the result that the two pictures will fuse into one image—in fact, we see just about what the giant would have seen. Some people can take the two pictures and fuse them without a stereoscope. This is a useful trick, not very difficult to learn.

Exercise 10. Calculate the difference in parallax between the top and the bottom of this mountain. H 5000 ft., h 1000 ft., B 1000 ft., take f as 153 mm.

$$\text{P of top} = 153 \times 1000/(5000 - 1000) = 38\cdot25 \text{ mm.}$$
$$\text{P of bottom} = 153 \times 1000/(5000 - 0) = 30\cdot60 \text{ mm.}$$

$$\text{Difference} \quad 7\cdot65 \text{ mm.}$$

THE STEREOSCOPE. The stereoscope was, curiously enough, invented before photography—and many centuries before that Euclid clearly understood that each eye perceives from a different viewpoint, thus receiving a

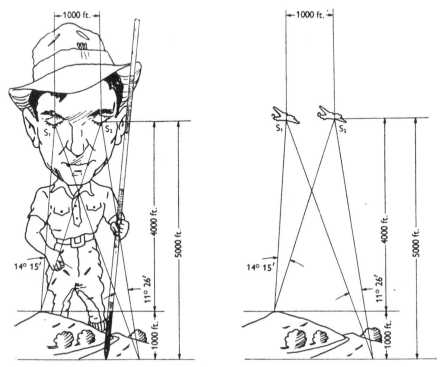

Fig. 34. The giant and the mountain.

slightly different image. Two types in common use are illustrated in figs. 35 and 36. Fig. 35 is a stereoscope consisting of two plano-convex lenses of about 90 mm. focal length. Each eye sees only one photograph—or a portion of it—and the magnification is of considerable advantage in interpretation.*

Fig. 36 is a mirror stereoscope of modern design. It consists of two pairs of mirrors, or the equivalent, arranged as shown diagrammatically in fig. 37. The angle of the mirrors is 45° and, in the best instruments, they are front surfaced. The front surfacing is advisable both because of the increased reflexion so obtained and because of the absence of double images. Prisms may be substituted for the small mirrors.

Figs. 37 and 38 show how the instrument works, and how the image is formed. An interesting discussion of stereoscopy in general, with many instructive diagrams, is given by Hart in (9), Chapter VI.

Exercise 11. Draw up a diagram such as fig. 38. Investigate what happens to the model when photo 2 is displaced to the right. The air-base and

* Lens stereoscopes should not be focused for infinity, the distance from the centre of the lens, fig. 35, to the picture should be sensibly less than the focal length. Some instruments require slight modification to this end.

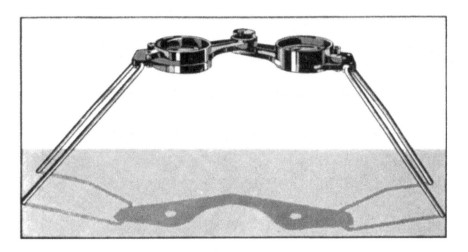

Fig. 35. Lens (or pocket) stereoscope. (By permission of Harrison C. Ryker, Inc., Oakland, Cal.)

Fig. 36. Stereoscope Universal folding, with parallax bar in use. (By permission of Messrs Casella Ltd., London.)

To face p. 64

eye-base remain unchanged as the photograph is shifted. Only the top part of the diagram need be redrawn in the new position.

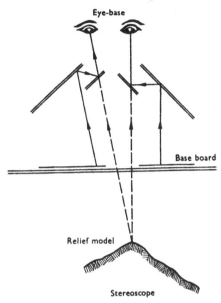

Fig. 37. Stereoscopic diagram.

FLOATING MARKS. In a stereoscope place a pair showing mountainous country. Pick a feature on the top of a mountain or hill and, on each photo, mark it with a small ink dot. Looking at the pair in the stereoscope, the two dots will, of course, appear as a single dot at the top of the mountain. Now if we have two bits of glass with a dot etched on each, and if we place these etched dots, instead of the ink dots, on the image positions of the feature, we see the same thing. Two dots fused to a single dot sitting on the top of the mountain. This is position 1 of fig. 39. Pieces of clear acetate, with scribed dots filled with chinagraph, may be used instead of glass.

Keeping the dots the same distance apart, move them to the position 2–2. The dots will still be fused into one; their parallax is unchanged, so they must appear as a single dot floating in space over the relief model at the elevation of the top of the mountain. This is position 2 (fig. 39).

Now, keeping the left mark still, move the right mark towards position 3. As the motion of the right mark is continued the fused image will slide down the line 2–3 and may be dropped to the ground at 3.

The dots now rest on exactly the same points of detail on each print, and the parallax will be that of the bottom of the mountain, namely, 30·60 mm., using the data of Exercise 10. The *difference* in parallax between the top and the bottom of the mountain is 7·65 mm., and this difference could be obtained by measuring the distance 2–3 on the right-hand photo.

Moving the floating marks, as they are called, by hand and measuring the difference in their positions would be little if any better than the centimetre-rule method, so we attach the pieces of glass to a rod and make provision to move one mark independently of the other, using a micrometer screw both to move the mark and to measure the motion. Such an apparatus is called a parallax bar.

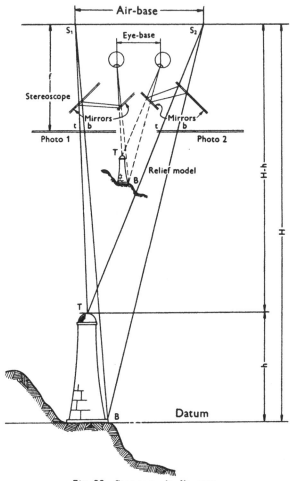

Fig. 38. Stereoscopic diagram.

DESCRIPTION OF THE PARALLAX BAR.   This simple precision instrument (fig. 40) consists essentially of two glass pieces attached to a rigid frame or bar. The right-hand glass is movable in the direction of the axis of the bar by means of a micrometer head upon which the motion may be read to $\frac{1}{100}$ mm. ($\frac{4}{10000}$ in.). The other glass may be moved axially a considerable distance—so that the one size of instrument may be used both for small and for large photographs, and in stereos of different sizes. The left-hand glass is rigidly clamped in the desired position and, once set, need not be changed working on the same size pictures in the same stereoscope.

It is seen that the instrument is used, not to measure the parallax of a point, but to measure differences in parallax between two points.

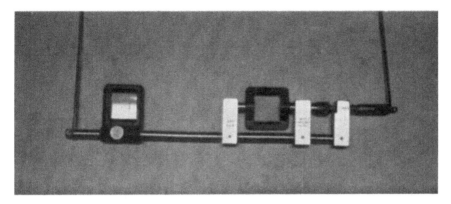

Fig. 40. Parallax bar. (By permission of Messrs Casella Ltd., London.)

Fig. 43. Simple stereocomparagraph. (By permission of Messrs Casella Ltd., London.)

To face p. 66

The etched marks on the glasses are found in a great variety of sizes, colours, and shapes to suit different conditions, and the whims of different photogrammetrists. There are often two or three sets of marks on the one instrument.

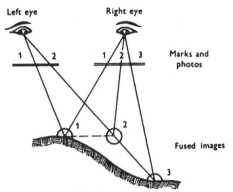

Fig. 39. Floating marks.

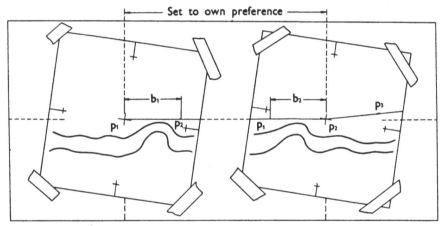

Whatman's paper, thin cardboard etc. fasten corners with cellulose tape.

Fig. 41. Method of mounting photographs for stereoscopic measurement of parallax.

Use of the Parallax Bar to determine Parallax Differences

MOUNTING THE PHOTOS AND SETTING THE BAR. In using the parallax bar to measure parallax differences, the photographs will first be base-lined, after which the preliminary procedure is as follows (fig. 41):

(i) Place a pair in the stereoscope that you are going to use and determine the separation of the principal points for comfortable accommodation.

(ii) Take a piece of Whatman's or other suitable stock and draw a centre-line through its length. Through the width of the mount draw two lines at right angles to the centre-line such that their distance apart is the required separation (from (i)) of the principal points. These cross-lines should be about equally spaced from the ends of the mount.

(iii) Attach the photos to the mount, their base-lines coincident with the longitudinal centre-line, and their principal points the desired distance apart as measured by the cross-lines. Use small tabs of cellulose tape at the corners when mounting.

(iv) Set the bar micrometer in the approximate centre of its run. Identify a prominent point of detail on each photo. Place the right-hand mark—the one moved by the micrometer—over the image on the right photo, unclamp the left mark and set it over the corresponding image point on the left photo, clamp the left mark and the instrument is ready for use. This operation is called zeroing the bar.

(v) Attach the mounted photos to the base of the stereoscope. Using stereos without a base, tape the mount to the drawing board or table and set the stereo over them. The two instruments are now ready for use.

(vi) It is advisable to arrange the mounted photos so that the light source is in the same direction as was the sun when the photographs were taken. (Keep the shadows towards you.) This aid is purely a psychological one, but the effect is remarkable.

READING THE BAR. Before attempting to determine elevations using the parallax bar it will be necessary to attain some proficiency in taking readings. The following exercise should be repeated until the specified precision is attained.

Exercise 12.   Readings with the parallax bar.

(i) You are given a pair—base-line them and, on the left photo only, mark at random a number of points of detail. The points will be marked with a small needle prick, and a large chinagraph circle may be drawn around the needle prick to mark it. A large circle is used so that stereoscopic impression near the point is not disturbed.

(ii) Mount the photos, zero the bar and place the photos under the stereo, fastening them down.

(iii) Place the bar on the photographs and fuse the marks. You will note that the bar should be parallel with the base-line, otherwise the marks will be difficult to fuse, and tend to do this (:). Gradually decrease the separation of the marks and the fused image will appear to rise above the ground. Increase the separation and they will float down towards the ground. Compare what you see with fig. 39. Pick out a hill-top, or a point on a side hill, and go through the motions of fig. 39, comparing what you see with the figure.

(iv) Bring the mark somewhat above the ground, drop it gently until it seems to go under. Repeat, watching carefully. Note the characteristic action as the point touches the ground.

(v) Repeat (iv), this time reading the vernier when you think the mark is on the ground. Record this reading and take four more readings. It is to be expected that, at first, these readings will differ. The differences should become smaller with practice. The required standard is that the maximum deviation of any one of five readings from their mean shall not exceed plus or minus 0·04 mm., and that the average differences, regardless of sign, should not exceed 0·03. The micrometer will not be read closer than the nearest hundredth of a millimetre. Use several points in this practice.

This is a preliminary training, not an operational, standard. Personnel who are to be trained as operators of automatic plotting instruments should have parallax bar acuity better than 0·01 mm.

(vi) Readings should be recorded and differences and means determined in the following form:

To be rejected		Satisfactory	
Readings	Diff.	Readings	Diff.
7·75	0·07	7·73	0·03
7·68	0·00	7·68	0·02
7·63	0·05	7·71	0·01
7·71	0·03	7·72	0·02
7·64	0·04	7·66	0·04
Total 38·41	0·19	38·50	0·12
Mean 7·68	0·038	7·70	0·024

Note that the third decimal place of the mean reading is not significant.

PARALLAX OF A POINT. In calculation of elevation from parallax readings we have a point of known elevation from which to start; this is called the datum point. We find its parallax from the parallax equation, $P = fB/(H-h)$. We then take readings with the bar on the known datum point and, as well, on the point or points whose elevation we wish to determine. The difference of parallax between the datum and an unknown point is obtained by subtracting the reading of the datum from that of the point. This difference in parallax between two points is denoted by the letter p.

Exercise 13. Using the points marked in Exercise 12, choose one of these as the datum and find p for the other points.

THE PARALLAX GRAPH. Knowing the elevation of a datum point, elevations could be determined by calculating from the parallax equation the parallax of the datum. The algebraic sum of this parallax and the measured parallax difference is the parallax of the unknown point. Its elevation could then be determined by substitution in the parallax equation. This procedure, which is rather tedious, is not necessary, as a solution may be effected by means of a graph.

There are to be expected quite appreciable differences in H and B, if not in a flight at least in a series covering a sheet. For most purposes it will be sufficient to take a mean height and a mean value of the air-base. Just what differences are permissible will be better understood with experience; for the present you will be given, or told how to obtain, the mean height and the mean air-base. Given these, and f, work of calculating the graph is as follows.

Exercise 14.   Calculate and draw a parallax graph.

(i) Data: f 3·25 in. = 82·55 mm.; H 6800 ft.; B 1200 ft.

(ii) Using the parallax equation, work out the parallax corresponding to points whose elevation is 0, 100, 200, 300, etc., to the greatest elevation you are likely to encounter.

Arrange a tabulation as follows:

h	(H-h)	P	
0	6800	14·57	As P is required to the nearest $\frac{1}{100}$ mm., a slide-rule cannot be
100	6700	14·79	used. P is obtained by dividing successive values of (H-h)
200	6600	15·01	into fB which, in this case, is 82·55 × 1200 = 99,060. Logs
300	6500	etc.	may be used, but even if H is not an even hundred, we may
400	6400		arrange that (H-h) is. Example, H = 6840; use values of
500	etc.		h of 40, 140, 240, etc., so that (H-h) is always an even
			hundred. In this way, long division is quite simple.

(iii) Using squared paper, plot P against h. You must be able to estimate the values of P from the curve to the hundredth of a millimetre, so choose a large scale. 1 in. to 100 ft. for the heights and 1 in. to 1 mm. for parallax is satisfactory. The bottom left-hand corner of the graph for the above data will be P 14·00 mm., and h zero. The curve must be clearly marked with the data for which it was made, i.e. with f, H and B. The date and the computer's name should also appear.

The use of such terms as absolute parallax, differential parallax, formula parallax and the like lead to confusion. If a man is 6 ft. 2 in. tall and his friend is 6 ft. 0 in., their heights are 6 ft. 2 in. and 6 ft. 0 in. respectively. The difference in their heights, referred to the tall man as the 'datum', is minus 0 ft. 2 in. and that is all there is to it—so with parallax. If the parallax of point A is 86·54 and of point B 81·52, the difference in parallax, referred to A as datum, is − 5·02 mm. *In the sign convention used in this book, a positive parallax difference always indicates a positive difference in height.*

CHAPTER 6

PARALLAX AND ELEVATION CALCULATIONS

Determination of Elevation Differences, Simple Case

THE BASIC CALCULATION. The principle of this calculation should now be understood. Having only one elevation on a photograph, procedure would be as follows:

(i) Enter the elevation of the datum in column 5 of the tabulation below.

(ii) Read its parallax from the parallax graph, P for h = 150 ft. is 53·64 mm. Enter this in column 4.

(iii) Take five readings with the bar on point 1, the datum. If the mean of these five readings is within the limits specified, enter this mean in column 2, opposite point 1. The limits are: mean difference not to exceed 0·03 mm., maximum difference not to exceed 0·04 mm. If the mean reading does not fall within these limits, repeat the whole series of five until satisfactory results are obtained.

(iv) Take the mean reading on point 2, and if within the limits of accuracy just specified, enter in column 2, opposite point 2.

(v) Determine the difference in parallax between the datum 1, and point 2 by subtracting the reading of the datum from that of the point. Enter this difference, p, in column 2.

Caution. Certain bars are graduated so that an increase in elevation results in a decrease in parallax reading. When using such a bar, subtract the reading of the point, *and change the sign of the remainder.* The practice of inverse bar graduation seems to result from considering depths below the air-base, rather than heights above datum.

The sample tabulation is for a Glauser bar, reading inversely. Thus reading at 2 is 12·50, subtract 12·25 (the datum reading) remainder, *plus* 0·25. Change the sign of this remainder, and p is *minus* 0·25. Similarly with point 3, where p is +2·23.

PARALLAX TABULATION (GLAUSER)

1 Point	2 Mean reading	3 P	4 P	5 Eleva- tion	6 Remarks
1	12·25	—	53·64	150	Datum
2	12·50	−0·25	53·39	132	
3	10·02	+2·23	55·87	308	

(vi) We have entered −0·25 in column 3; hence the parallax of point 2 is the algebraic sum of the parallax of the datum and the parallax difference. 53·64 − 0·25 = 53·39.

(vii) From the graph the elevation of a point whose parallax is 53·39 is 132 ft. Enter this under elevation, column 5.

(viii) Similarly with point 3 and succeeding points, p for 3 is 10·02 − 12·35 = − 2·25. Changing the sign, we have + 2·23 in column 3, whence P = 53·64 + 2·23 = 55·87, following the same procedure as before. The elevation from the graph is then 308 ft.

Exercise 15. Using the above procedure, determine the elevation of the points you marked in Exercise 11, and compare these values with the elevations shown on the map.

Particularly if the points are some distance apart on the picture and at any considerable difference in elevation, this exercise will bring out that uncorrected parallax readings cannot be relied upon to give accurate elevation differences.

CORRECTION. If we wish to make use of parallax readings to obtain elevations it remains then to devise some method of correcting these computations. For the simple correction drill given the principle involved is that on a line at right angles to the base-line, error varies directly as the distance from the datum. Thus, if we confine our parallax readings to such a line, and have a second point of known elevation, a tie point, at the end of this line, we can get the correction at the tie point quite easily. The correction at other points may then be done by a simple graphical construction.

An analogy will help to visualize this; on obliques the horizontal scale is constant on a line at right angles to the principal vertical, which you may compare to the base-line. As you move in the direction of the principal vertical however, the scale changes rapidly.

Standard Heighting Drill, Right Angles to Base-line

DRILL. This can be followed by working through an example. Fig. 42 shows Parallax Form 1, to be used for elevation calculations where the unknown points are at right angles to the base-line, and where there is a point of known elevation at or near each end of the line. Drill is as follows:

(i) The datum point 1, elevation 195 ft., and the tie point 5, elevation 376 ft., are given. The points 2, 3 and 4 are chosen to lie on a line at right angles to the base-line and represent topographic features whose elevation we wish to determine.

(ii) Having set up the photographs as previously explained, take readings on the datum and enter the mean in column 2 of the form. Now take readings on the tie point 5, entering the mean in the same column opposite 5.

(iii) Column 3 is the uncorrected parallax difference at the tie and is equal to $3\cdot16-4\cdot28=-1\cdot12$, but since we are using the Glauser bar the sign is changed to plus.

(iv) From the graph, enter in column 6 the parallax of the datum and of the tie, $54\cdot26$ and $56\cdot95$ respectively.

Reproduced by permission of R.A.F.

AIRSURVEY

1 Cdn Fd Svy Coy R C E

PARALLAX FORM I

Observer *HCH* Date JUL 18 1943 Flight *AC-54*

Computer *CF* Date JUL 18 1943 Photo No *22-3*

f *82.55* mm H *4000* ft B *2500* ft

To obtain p for any point, SUBTRACT the reading of the datum from that of the point. When using the Glauser parallax bar CHANGE THE SIGN OF THE REMAINDER, when using other bars, or the ZD-15, the sign is NOT changed.

Correction at tie = P of tie - P of datum - p of tie. The correction of intermediate points varies as their distance from the datum, scaled from the photo. These corrections are taken off the correction triangle.

Glauser Bar Figure 5

Point	Micrometer or vernier reading	p	Corrn	p	P	h	Remarks
1	4.28	-	-		54.26	195	Datum
2	6.12	-1.84	+0.96	-0.88	53.38	132	Valley
3	4.29	-.01	+1.22	+1.21	55.47	280	
4	2.06	+2.22	+1.40	+3.62	57.88	435	Top of Hill
5	3.16	+1.12	+1.57	+2.69	56.95	376	Tie

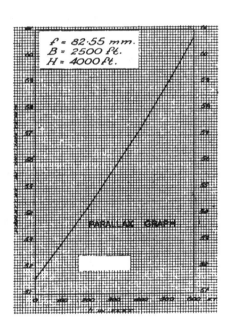

$f = 82\cdot55\ mm.$
$B = 2500\ ft.$
$H = 4000\ ft.$

PARALLAX GRAPH

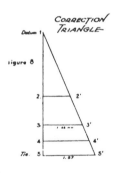

CORRECTION TRIANGLE

Figure 6

Fig. 42. Parallax heighting drill, right angles to the base-line.

(v) P of tie $-$ P of datum $=56\cdot95-54\cdot26=2\cdot69$, and this should be the difference in parallax between the tie and the datum. But we measured the difference with a parallax bar and got only $1\cdot12$ mm. difference. This discrepancy results from any of the circumstances which would cause want of correspondence and, in addition, from inexact alinement of the photographic base-lines. In order to correct this measured parallax we must then add 1·57 mm. to make the corrected parallax 2·69 (2·69 − 1·12 = 1·57). The correction is obtained as follows. Correction at tie = P of tie − P of datum − p of tie. This rule is printed on the form. Enter the correction, + 1·57, in column 4 opposite point 5.

(vi) To find the correction at any other point, draw the correction triangle. Base 5–5′ is equal to the correction, 1·57 mm., drawn to the scale 1 in. equals 1 mm., and height 1–5 is equal to the distance 1–5 on the photo. Points 2, 3 and 4 are transferred with dividers, or with a strip of paper, from the photograph to the triangle. The correction for intermediate points may now be obtained from this correction triangle; thus the distance 3–3′, representing 1·22 mm., is the correction at 3. Enter the correction for each point in the correction column.

(vii) Complete column 3 by subtracting the datum readings successively from the point readings and changing the sign.

(viii) The corrected parallax difference, p, column 5, is the algebraic sum of p′ and the correction; enter in the proper column.

(ix) Column 6 is the algebraic sum of the P of the datum and the corrected parallax difference p.

(x) Elevations corresponding to these P's are then read from the graph.

The necessity of keeping at right angles to the base-line is stressed for the reason that, as this condition is departed from, heighting errors increase rapidly if tilts are large, and that these errors cannot be corrected by a simple straight-line adjustment. If tilts are within 3 or 4, perhaps 5°, certainly the siting of points a few millimetres off will do no harm—but if one adheres strictly to the condition there is no need to worry about tilt.

Exercise 16. Given a pair of photographs and a 1:25,000 map, choose datum and tie points at right angles to the base-line and, using the procedure just described, determine the elevation of a number of intermediate points. Compare your values with the map elevations, and find the average difference between the photogrammetric elevation and the map elevations expressing this difference as a constant times H/b.

Contouring

How Photogrammetric Spot Heights are Used. Given two points of ground control, the line joining which is substantially at right angles to the base-line, it is seen how additional points on this line may be heighted photogrammetrically.

Given several pairs of such spot heights, a net of levels may be obtained across the overlap, and from these the photograph may be contoured, proceeding in much the same way as when drawing contours from a grid of spot heights obtained from ground observations. There is, however, the great difference, and advantage, that the stereoscopic model shows the actual shape of the ground, and that accordingly the true shape of the contour may be depicted quite faithfully.

INSTRUMENTAL CONTOURING. Instruments, both simple and complex, are available to effect the actual drawing of contours. The simple type (fig. 43) consists of a parallax bar and tracing pencil with means to maintain the bar, as it is moved over the model, parallel to the base-line. The vernier is set to a particular elevation, the floating mark placed in contact with the ground by shifting the bar, and the pencil dropped. The floating mark may now be caused to follow a contour by moving it around the relief model, always keeping the mark in contact with the ground. Corrections are required for tilt. Note that such simple devices only plot a perspective projection of the contour, as it would appear on one of the photographs of the pair. All contours are taken off at the same scale. Muzzafer Tugal (12) describes a design in which the motion of the micrometer screw, as it is changed to trace a contour at a different elevation, makes a proportionate change in the scale of the trace, thus producing an orthogonal projection.

The more complex instruments (actually they are simple in principle) may be used for extending vertical control, bridging from one controlled model to another at a considerable distance. As well, they take off a true plan of each contour and of detail, rather than a perspective projection.

Correction by Observation of K

ERROR IN CARRYING ELEVATIONS FORWARD. The effects of tilt and of inclination of the air-base are capable of rigid mathematical analysis. If an instrument of precision, such as the Thompson stereocomparator, be used to measure parallaxes and want of correspondence on glass negatives, or on foil prints, elevations may be carried across several overlaps. But the validity of the result is so readily upset by minute inaccuracies that the heights carried across several overlaps are subject to discrepancies. In addition, the formulas are so complex, and take so long to apply, that the tendency is to provide enough heights from ground survey for the simple, right angle, method to be used. This requires dense ground control and is therefore costly. The alternative is to use automatic plotting apparatus which not only requires much less ground control, but also turns out work of higher grade.

In passing it is pointed out that the instrument does not exist in which photographs are fed into one end, and out of the other end comes a contoured map with spot heights to a tenth of a metre, without intervention of ground control or human agency.

In spite of the complexity of the application of the mathematical analysis (see (10) and (7)), certain basic principles may be examined in general terms. Let us consider an exactly vertical pair (fig. 44) exposed from two camera stations the same height above datum. The distance of any image point from the base-line must be the same on each photo, and must correspond.

CAUSES OF K. On fig. 44, A (which is any ground point whatsoever) appears at a_1 on photo 1 and at a_2 on photo 2. Then, if the pair be exactly vertical and if S_1 and S_2 be at the same height above datum, a_1 and a_2 will be at the same distance from their respective base-lines. To make this quite clear examine fig. 45, which is a front elevation of fig. 44, looking back along

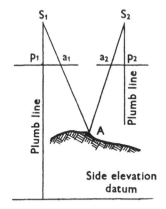

Fig. 44. Vertical pair, side elevation.   Fig. 45. Vertical pair, front elevation.

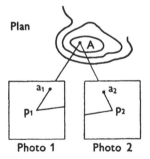

Fig. 46. Vertical pair, plan.

the line of flight from some position ahead. S_1 and S_2 are at the same height, and will therefore appear as one in this view. The positive photo-planes will all coincide. A is any point of ground detail, and the rays from A to S_1 and S_2 (which look like a single ray) will appear to cut the photo-plane in a single point a_1a_2; hence their y ordinates in fig. 46 will be equal.

Exercise 17. Under what other conditions, short of true verticality, will there be no want of correspondence? Equal lateral tilt of both photographs. Draw a diagram like fig. 45 to illustrate this. (Just tilt the datum line, and mark the plumb-line 'optical axis'.)

In general, then, when both photographs are in the same plane, there will be no want of correspondence. Hence, if we look through a stereoscope and the ground appears to slope rather uniformly, in the absence of ground control, or knowledge of the terrain, we cannot tell whether the ground does in fact have this slope, or whether both photographs are equally tilted.

It is thus evident that want of correspondence measurements alone cannot determine whether a pair is vertical or not, and therefore that if we correct the effects of want of correspondence we only put both photographs in the same plane. The complete solution requires a knowledge of the true elevation of certain picture points. That is to say, we may determine relative tilts, but must have control to find absolute tilt.

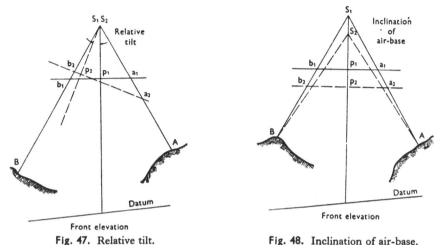

Fig. 47. Relative tilt.   **Fig. 48.** Inclination of air-base.

EFFECT OF TILTS. The effects of these relative tilts can be very well shown on the multiplex, and this demonstration should be made to personnel under air-survey instruction. In the absence of distortion, all want of correspondence results from the following simple tilts, either singly or in combination with one or more of the others: relative lateral tilt, inclination of the air-base, relative fore-and-aft tilt.

Suppose that photo 2 is tilted as shown in fig. 47; this is relative lateral tilt. a_2 is farther away from the base-line than is a_1. Note that a point B, on the opposite side of the base-line, will show want of correspondence of opposite signs.

In fig. 48, S_2 is lower than S_1, the air-base is inclined. Both A and B will show want of correspondence, this time in the same direction, simply a scale change.

Now consider relative fore-and-aft tilt and let the back of photo 2, the part containing a_2 in fig. 49, be tilted down, while the forward part tilts up. Here a_2 and b_2 will both be farther away from the base-line if they are in the position shown in the figure. What would happen to points to the right of p_2 on photo 2? Such points would move closer to the base-line. Contrast this with the effect of inclination of the air-base.

There can be no difficulty in understanding that these changes in correspondence will occur as described, and in realizing that the mathematics relating the changes in correspondence to tilt, however complex, may be worked out. It is also evident that the changes in position resulting from tilts will also be associated with changes in parallax, since the tilt displacement will be radial from the isocentre and thus have an x component. So if we take measurements at scattered points on tilted photographs, before we use them to calculate heights, we must correct them. The correction can be calculated from measurements of want of correspondence; then we may proceed in much the same manner as before, tieing in to ground control.

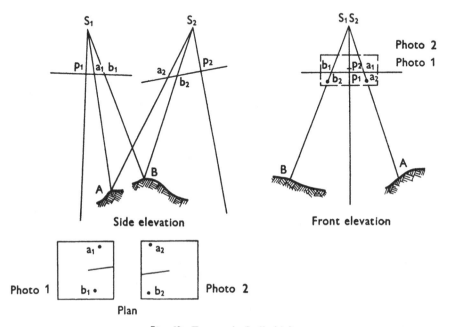

Fig. 49. Fore-and-aft tilt (tip).

GENERAL DESCRIPTION OF THE ZD-15 STEREOSCOPE. So far we have measured K merely with a good rule giving its value, in skilled hands, to one- or two-tenths of a mm. The ZD-15 topographic stereoscope, manufactured for the War Office by Barr and Stroud, is not now in general use, but it will be of value to describe the instrument, as the method of measuring K is of interest.

An essential feature is a pair of grid plates, a half-size reproduction of the markings on one of which is shown in fig. 50. Consider a pair of prints set up in an ordinary stereoscope, in precise alinement. Let a glass grid, such as fig. 50, be placed over each photograph, the centre row of crosses accurately coincident with the base-lines in each case.

Obviously we can fuse the one grid with the other, and so fused this stereoscopic image will define a horizontal plane. If the left grid be fixed and the right grid be moved towards it, the fused plane will rise; increasing the separation will cause the plane to fall. The relative motions of the two grids may be measured to 0·01 mm. by means of a vernier, and we can obtain parallax differences just as with a parallax bar. The vernier reads so that increase in parallax indicates increase in elevation. Provision is also made to move both grids together without altering their separation. The instrument will also measure the x and y co-ordinates of any image point.

ZD–15 grid

Fig. 50. ZD-15 grid.

MEASURING K WITH THE GRIDS. Consider the ground point A, the image position of which is a_1 on photo 1 and a_2 on photo 2, fig. 51. Let a grid cross of the left-hand grid be exactly over a_1. When there is want of correspondence, as is illustrated, the right-hand cross, the intersection of the two arms, cannot be placed exactly over a_2, so that the intersection itself cannot be set on the ground at a. But it will be possible to bring the N.E. arm to the ground at the stereoscopic position of the fused image a, and the N.W. arm of the cross must then appear to be floating in space above. This, is the characteristic want of correspondence phenomenon, i.e. the two arms of the cross not being coplanar. The position of the right-hand cross will be as illustrated by the solid lines of fig. 51 (see also fig. 74).

It will be possible, by moving the right-hand grid to the position shown by the broken lines, to bring the N.W. arm to the ground, and under these circumstances the N.E. arm will appear below the ground. Since the crosses are at $45°$, the difference between successive positions of the two arms is $2K$ as shown. Hence $K = \frac{1}{2}$(N.E.–N.W.).

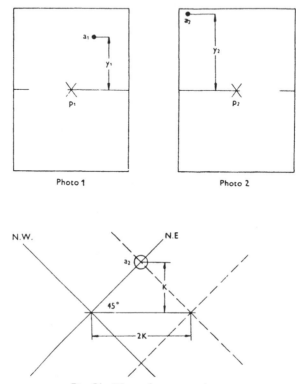

Fig. 51. Want of correspondence.

BASE-LINING WITH THE ZD-15. The correspondence principle is used to base-line in ZD type stereoscopes. By definition, K is zero along the base-line. Hence if K appear at any point on the base-line, as evidenced by the arms of the crosses on the base-line not being coplanar, the bases are not in alinement. The want of correspondence is removed by rotating the turn-tables upon which each photograph is mounted. Remove at p_1 by rotating 2, and at p_2 by rotating 1. When this adjustment is completed, the base-lines are drawn in the instrument.

K BY PARALLAX BAR. K can be measured with a parallax bar if one rotates each photograph $90°$ in a counter-clockwise direction and sets them in a stereoscope, as shown in fig. 52. The base-lines should be set as nearly parallel as can be done by ordinary means. With the photographs so set, hills and valleys will not appear in relief, but want of correspondence at a particular point, if present, will be equivalent to a parallax difference in the ordinary setting and so give the impression of relief.

TAKING THE READINGS. Take a reading at the point, and a second reading on the base-line at the foot of the perpendicular from the point. The difference between these is the required K. With inverse reading bars, K = reading at point − reading at base-line. For direct reading bars, K = reading at base-line − reading at point. This is in accordance with the definition $K = y_1 - y_2$.

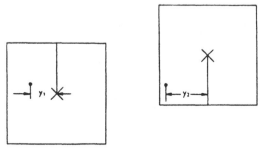

Fig. 52. Measurement of K with the parallax bar.

Obviously if the base-line be inaccurate, K will be inaccurate too. Fortunately, an error in base-lining will itself show up as want of correspondence and the error may be corrected. If inaccurate, the fused base-lines will not lie on the ground, and probably will seem to slope.

These parallax differences are small, therefore more than usual care is required with the readings. This method is of a lower order of precision than the ZD-15, for no matter how skilled one is, it is very doubtful whether it is possible to draw two lines, each to be within, say, four ten-thousandths of an inch of its true position, that is, within 0·01 mm. (Such small inaccuracies cannot show up as relief, since the line itself will obscure detail for a greater width than this.) The ZD-15 can set the photographs to somewhere near these limits on the base-line, although one cannot draw the base-lines so closely in the instrument.

Parallax Scale

SLOPE OF PARALLAX CURVE. We have shown that

$$P = \frac{fB}{(H-h)}.$$

Differentiating with respect to h, we have

$$dP = \frac{fB}{(H-h)^2} dh.$$

Whence

$$\frac{dh}{dP} = \frac{(H-h)^2}{fB}.$$

In some works P is differentiated with respect to (H-h), rather than h, thus obtaining a minus sign in the above expression. Since parallax increases as (H-h) decreases, we have the argument for inversely graduated parallax bars, and similar instruments. As in determining ground elevations we are concerned with h, not H-h, the direct procedure seems more reasonable.

dh/dP is ft./mm., and may thus be looked upon as the parallax scale. For example, where $H = 4000$ ft., $h = 50$ ft., $f = 82\cdot55$ mm., $B = 2500$ ft.,

$$\mathrm{dh/dP} = \frac{(4000-50)^2}{82\cdot55 \times 2500} = 75\cdot7 \text{ ft./mm.}$$

The parallax scale is not constant over the whole of the photograph, nor is the horizontal scale. Both vary with (H-h). But just as we refer to the (horizontal) scale of a vertical photograph taken under certain conditions of f and (H-h), so we may refer to the parallax scale—implying an average value.

Exercise 18. Compare the value of dh/dP obtained above with that given by the parallax curve of fig. 42.

From the curve　　　　$h = 100$　$P = 52\cdot91$

　　　　　　　　　　　$h = 0$　$P = 51\cdot59$

　　　　　　　　　　　$\overline{}$　$\overline{}$

　　　　　　　　　　　100 ft. $= 1\cdot32$ mm.

Therefore　　　　　　1 mm. $= 100/1\cdot32 = 75\cdot7$ ft.

SIMPLIFIED SLOPE FORMULA. dh/dP can be found more simply as under:

$$B = b\frac{(H-h)}{f}.$$

Substituting　　　　$\dfrac{\mathrm{dh}}{\mathrm{dP}} = \dfrac{(H-h)(H-h)}{f \times \dfrac{b(H-h)}{f}} = \dfrac{H-h}{b}.$

Exercise 19. Compare the short-formula result with that obtained in the preceding exercise, b measured $52\cdot2$ mm.,

$$\mathrm{dh/dP} = 3950/52\cdot2 = 75\cdot7 \text{ ft./mm.}$$

RESTRICTION OF APPLICATION OF SIMPLIFIED FORMULA. There is a reservation with regard to the use of this short formula. The h in (H-h) ×(H-h) of the numerator is the elevation above datum of the feature whose height we are measuring. The h in (H-h) of the denominator is the h used to determine the photo-scale of the base-line b, i.e. the elevation of p_2 if we use photo 1 on which to measure b, or the elevation of p_1 if b is taken from photo 2. These two h's are not necessarily the same. Either p_1 or p_2 will, ordinarily, be sufficiently close to the elevation of the feature so that no error, commensurate with the method, is introduced. Measure b on the photograph, the elevation of the principal point of which differs more from that of the feature.

When the difference between the elevation of the feature and that of the principal point closer to it in height is estimated to exceed 5 % of H, the longer formula should be used. Table 9 illustrates the magnitude of the error.

TABLE 9.   *Error of simplified parallax formula*

H = 4000 ft.   B = 2500 ft.   b = 51·7 mm.   f = 82·55 mm.

Difference in elevation between feature and relevant principal point, percentage of H	Error in feet caused by using short formula, measuring a height of about 25 ft.
1%	0·2
2	0·4
3	0·7
4	0·9
5	1·1
6	1·3
7	1·5
8	1·7
9	1·9
10% of H	2·1 ft.

USE OF $dh/dP = H\text{-}h/b$.   Examples of the use of the short formula are given in the following exercises:

Exercise 20.   (H-h) 7000 ft., b = 50 mm. The difference in parallax between the top and the bottom of a cliff is 0·62 mm.; find the height of the cliff.

$$dh/dP = 7000/50 = 140 \text{ ft./mm.}$$

Hence $$h = 0·62 \times 140 = 87 \text{ ft.}$$

Exercise 21.   With a camera giving 9 × 9 in. contact prints, what is the greatest height an aircraft can fly to get photogrammetric elevations to within 10 ft.? Assume that you can rely on parallax differences to within 0·03 mm. and that other sources of error amount to a further 0·02 mm.

We require that 0·03 + 0·02 be equivalent to 10 ft. On a 9 × 9 in. print with standard 60% overlap, b = 0·4 × 9 = 3·6 in. = 90 mm. (about).

Hence since 0·05 mm. = 10 ft., 1 mm. = 10/0·05 = 200 ft./mm.

$dh/dP = H/b = 200$ ft./mm., whence $H = 90 \times 200 = 18,000$ ft.

Exercise 22.   Given 5 × 5 in. contact prints exposed from 30,000 ft., could you locate 10 m. contour crossings on a line at right angles to the base-line, given tie and datum elevations? Assume accuracy equivalent to 0·05 mm. of parallax.   b = 50 mm., dh/dP = 30,000/50 = 600 ft./mm.;

accuracy = 600 × 0·05 = 30 ft., or about 10 m.

You could not, therefore, differentiate between the absolute positions of successive 10 m. contours by means of parallax bar readings.

Exercise 23.   Same data as previous question, except that format is 9 × 9 in., b = 0·4 × 9 × 25·4 = 90 mm.

$$dh/dP = 30,000/90 = 333 \text{ ft./mm.; accuracy} = \frac{333 \times 0·05}{3·28} = 5 \text{ m.}$$

This is better, but the photogrammetric contour positions may be displaced by one-half their distance apart.

WIDE RIVER FOR CONTROL. The examples of the use of dh/dP for finding h have, so far, been of points whose map, or plan position, is nearly the same, as, for example, a cliff. A common problem is to determine the profile at a proposed river crossing; here elevations are usually required for some distance either side of the river, and uncorrected parallax readings could not be relied upon.

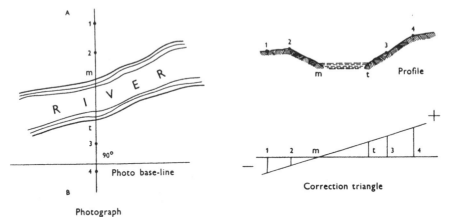

Fig. 53. Use of a wide river for control.

However, the river crossing can be heighted very well if the river appears reasonably wide in the photographs. The water-level at both sides of the proposed crossing is the same, and use can be made of this fact to control the raw parallax differences.

The typical case of fig. 53 can be solved by the **dh/dP** formula, and without ground control. It is required to obtain the section **AB**, where **AB** is at right angles to the base-line.

$$H = 5600 \text{ ft., } b = 92 \text{ mm., } dh/dP = 60\cdot8 \text{ ft./mm.}$$

Point	Reading	p'	Correction	p	h = 60·8 × p
1	6·25	0·75	−0·20	0·55	33 ft. above water-level
2	6·28	0·72	−0·10	0·62	38
m	7·00	Zero	Zero	Zero	0 datum, water-level
t	7·20	−0·20	+0·20	0·00	0 tie, water-level
3	6·90	0·10	+0·25	0·35	21
4	6·10	0·90	+0·35	1·25	76

Glauser bar

b is measured on the photograph, the elevation of whose principal point differs more from that of the feature.

The calculation should be followed without difficulty. The water-level at each side of the river determines the correction—these points are used as tie and datum. The parallax at **t** is the same as at **m**. Readings differ in the example by 0·20 mm. (from the usual causes); hence the correction at **t** is +0·20 mm. The correction diagram is set up in the ordinary way as shown. The points on the section should be pricked stereoscopically, using magnification, and the horizontal distances must be measured very carefully.

If it is desired to obtain data for a contoured plan, or if the required section is not at right angles to the base-line, several sections are taken parallel to **AB**.

The small difference in water-levels at the different sections is neglected. The same procedure must be gone through in each case, and a new correction triangle set up for each section. Not only will the river widths vary at the different sections, but also the observed parallax differences at shore level may not be the same.

This problem is a classical military one, and the above solution is satisfactory if low-altitude photography is operationally practicable, which rarely is the case. The probable heighting error is of the order of 0·08 (H-h)/b. Usually altimeter error may be neglected, as the measured height differences are small in comparison to H-h, but one must have information to determine height of aircraft above ground. The grid and analyser system is preferred as being more reliable.

A source of error not mentioned so far is that which arises when, due to dense undergrowth and the like, you cannot see the ground to float the mark down to it. This is sometimes called the pointing error and may be serious. The best thing to do is to make an allowance based upon your knowledge of the country—you may be able to get a pretty good idea of the height of such obstructions to your view from carefully examining the photograph.

Observational and other Errors and their Magnitude

ERRORS OF PARALLAX BAR HEIGHTING. A good observer can read parallax with an average error of 0·02 mm. and, since the mean of several readings is taken, the error of the mean should be about 0·01 mm. This is an observational error. But the differences in photographic x ordinates do not truly represent elevation differences because of distortion, image movement and so on. Further errors are introduced by inexact tilt correction and by the use of mean values of B and H. It should be remembered, too, that we are dealing with differences, and that the error of a difference between two observations each of probable error e is $\sqrt{(e^2 + e^2)} = 1·4$ e.

Exact mathematical computation is unjustified, but extensive ground checks indicate a total error equivalent to about 0·06 mm. working at right angles to the base-line on ordinary good photographs, with tilts of 2 or 3°, and with relief up to about 5%. In these tests ground data were spot heights and contours from the British 1/25,000 sheets (G.S., G.S. 3906).

RELIABILITY IN FEET. It has been shown that dh/dP = H/b. From this equation, and from the data of the preceding paragraph, the reliabilities shown in table 10 have been compiled.

Survey Analogy to Heighting Problem

EQUIVALENT SURVEY OPERATION. What we do photogrammetrically is measure a height difference by angular observation from each end of an air-base of known length. Consider the analogy of an equivalent ground-base running east and west, where the problem is to determine differences in

TABLE 10.   *Parallax bar heighting error in feet, for different altitudes and formats*

The table gives probable spot height error in feet for different altitudes and formats. Average error is shown, that is to say, in a series of spot-height determinations *half* of the errors may be expected to exceed the figures given.   The error is that to be expected running photogrammetric levels between a datum and a tie, each of known elevation, the line joining these points being at right angles to the base-line, and the relief being low.

Height of aircraft (ft. above ground)	Error in feet for format and base shown		
	5 in. b 50 mm.	7 in. b 70 mm.	9 in. b 90 mm.
2,000	2	2	1
4,000	5	3	3
6,000	7	5	4
8,000	9	7	5
10,000	12	8	7
15,000	18	13	10
20,000	24	17	13
25,000	30	21	17
30,000	35	25	20
35,000	41	29	23
Error, expressed as a fraction of H	H/850	H/1200	H/1500

Error is taken as $0.06H \cdot b$, implying good sharp photography. To attain these results points heighted must be well defined, and the distance from datum to tie restricted to about one-quarter of b. Prints should be waterproof for dimensional stability.

northing between distant points by means of angular observations at each end of the base-line. The easting and westing of the observed points lies between the easting and westing of the ends of the base. The following considerations will affect the accuracy of the ground-survey result:

(i) Length of base-line.

(ii) Distance to the objects whose difference in northing we wish to measure.

(iii) Precision of the device with which we measure the angular differences.

(iv) The pointing error, whether the object be well defined, say marked with a picquet, or otherwise.

PHOTOGRAMMETRIC FACTORS.   (i) and (ii) above are represented, photogrammetrically, by length of air-base and height of aircraft above ground, the ratio H/B depending upon the angle of the lens.

Considering methods which depend upon the (stereoscopic) measurement of photo-co-ordinates, the precision of co-ordinate measurement will not vary except in so far as definition may be better in one case than in another. Therefore the angular precision will be greater in the case of a long focal length cone than in the case of the short one. This is equivalent to saying that we get better results from a 6 in. 9×9 than from a $3\frac{1}{4}$ in. 5×5 (the angle is the same in both cases, both are very wide-angle lenses) as reflected in the formula $dh/dP = H/b$, since b is 90 mm. as opposed to 50 mm. in the 5×5.

Do we then get better results from a 12 in. cone than from a 6 in. cone, the format and altitude being the same? No—other things being equal—because considering the survey analogy, although we increase the precision of angular measurement, we decrease the air-base in the same proportion. Note, however, that the distortion and resolution characteristics of a long focal length lens may be better than those of a short focal length lens on the same format, so that—even though the parallax scale does not change—we may expect better definition and, accordingly, somewhat better results with a parallax bar.

For example, during the war it was found that better heighting was possible from 36 in. photography than from 6 or 12 in. photography from the same height, and on the same format. The writer takes the view that, format size being constant, it is height of aircraft above the ground which chiefly determines inherent precision—notwithstanding differences in height over air-base ratio.

Some authorities stress the height over air-base ratio, neglecting the effect of precision of angular measurement, i.e. angle subtended at S by 0·01 mm. on the focal plane. In order to appreciate the photogrammetric advantage of a wide field, compare, a wide-angle lens with a narrow-angle lens *of the same focal length*.

RELATIVE PRECISION OF DIFFERENT PHOTOGRAMMETRIC INSTRUMENTS. Considering instruments where the height difference is not deduced from co-ordinate measurement, the multiplex projector and the stereo-planigraph for example, the principles of the survey analogy apply. Working from the same photography it is seen that any comparison is simply that of the effectiveness of each apparatus considered as a device for measuring small angular differences, factor (iii) above being the only one which changes. One may thus judge for one's self the relative merits of such instruments with regard to topographic precision.

It is well, too, to remember that the skill and stereoscopic acuity of the observer is a considerable factor, whether he is working with a simple parallax bar or with a complex photogrammetric instrument.

Focal Length and Parallax, Model Effects

RELATION TO PARALLAX. The equation $dh/dP = H/b$ states that parallax is independent of focal length, height of aircraft and photo air-base (i.e. format) being constant. The effect is illustrated by fig. 54.

Exercise 24. Draw a diagram as fig. 54 enlarged two or three times. From this diagram prove that, although scale increases (horizontal scale) with increase in focal length, parallax remains constant, since, to maintain overlap at the enlarged scale, the air-base must be reduced.

EFFECT ON THE STEREOSCOPIC MODEL. The effect upon the relief model seen in an ordinary stereoscope is sometimes misunderstood. It is simply that although the horizontal scale changes when f is increased, the vertical scale does not. Fig. 55 shows this. The points a and b of fig. 55 or the top and bottom of the lighthouse of fig. 54 subtend, in each case, the same parallax *difference*. The horizontal scale of the model does, however, change in the manner indicated by fig. 55.

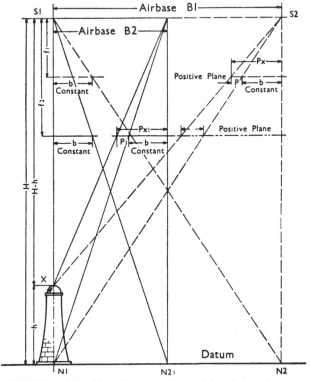

Fig. 54. Effect of focal length on parallax, height of aircraft constant.

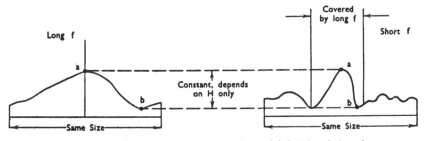

Fig. 55. Effect of focal length on the stereoscopic model, height of aircraft constant.

In sketching topography off contact prints, in rather flat country the contours could be traced much better on the short focus prints, since the relief is accentuated. If the country were very steep indeed, one might prefer the longer focal length. There is obviously an optimum photo-scale, but where we are concerned only with parallax measurement of height, it is not very critical. The horizontal scale must be so large that the points or features you wish to height are clearly discernible, but it is the distance of the air-survey camera from the ground that matters, rather than the angle of the cone (format size being constant).

EFFECT OF VIEWING DISTANCE. This effect is due to the viewing distance—the length of the path of light from the eye to the photograph—being dissimilar to the focal length of the air-camera lens. Only when this distance equals f, and when the ray from each eye to p of each picture is normal thereto, are the original perspective conditions recovered—and only in these circumstances is the visual model without distortion.

Where f is not recovered, but the eye is on the normal produced, and where there is no tilt, then the effect is simply a proportionate change in the ratio of vertical to horizontal scale.

Whether or not f is recovered, and whether or not the eye is on the normal, does not invalidate any techniques dependent upon photo-co-ordinate measurement (i.e. parallax bar), but some 'simple' stereo plotting instruments which make use of measurement of the virtual image model itself (formed as in an ordinary stereoscope) may be subject to gross error from these causes.

SETTING PHOTOGRAPHS IN A STEREOSCOPE—RECOVERY OF ORIGINAL PERSPECTIVE CONDITIONS. On p. 67 a drill was given for mounting photographs for the stereoscope, and the above points were not stressed, for the reason that—since we were measuring co-ordinate differences—comfortable accommodation is the important factor. However, when we wish to mark, say, the drainage pattern, model distortion may be of such magnitude as to make this difficult, particularly near heights of land, for uphill may very well appear as being downhill.

The photographs may be set in their proper positions by determining where the ray from each eye strikes normal to the viewing surface. This may be done with the aid of a small set square—when its vertical edge cannot be seen, the ray is normal. This point is determined for each eye and marked. The markings on the mount are similar to fig. 41, but the separation of the principal points is determined as above. However, when so mounted many people are unable, comfortably, to fuse the model.

ALTERNATIVE SETTING MEANS. An alternative means consists in setting the photographs at 90°, as fig. 52, and separating them until the area in question appears flat—the model will appear dished, flat, then pin-cushioned as the separation is altered. The above will give the answer over (comparatively) small areas, and it is important that this be done when, for example, interpreting height of land drainage from 6 in. contact prints in a stereoscope with a viewing distance of 10 or 12 in.

ENLARGEMENT AND MAGNIFICATION. If we work from an enlargement, rather than from a contact print, the focal length of the enlargement is greater in the direct ratio of the enlargement; therefore the horizontal scale of the model is enlarged in the same measure. The photo air-base **b** is increased as well, and in the same proportion. The result is thus equal enlargement of the stereoscopic model both in horizontal and in vertical scale. One might therefore expect, using a parallax bar, to obtain double the precision with enlargement to two diameters. This will rarely be achieved in practice, and enlargement to two diameters should be regarded as the extreme limit where parallax measurements are to be made. Unless the original negative is very sharp indeed, it may be found that the loss of definition is such that the smaller scale contact print may be preferred for this purpose. Great care must be taken in such an enlargement that distortions are not introduced; a precision enlarger is essential.

It is advantageous to introduce magnification in the stereoscope, and this is provided for in many instruments. It is found that the parallax differences, read with a parallax bar, do not increase appreciably in accuracy with magnification—it is ordinarily about 0·005 mm., too small a gain to be of advantage. What does happen, however, is that the mark can be brought more readily, and with greater precision, on the exact spot it is desired to height. Magnification should not be so great as to destroy definition, nor should the optical system be so complex as to absorb much light.

Owing to restriction of the field of view, magnification is not used for a general examination of the area of the overlap nor, if contouring the photograph, to obtain general topographic shapes. 'Character' may be added to contour shapes using magnification.

Limitations of Photogrammetric Methods

VALUE OF KNOWLEDGE OF LIMIT. A knowledge of the limitations of photogrammetric heighting methods is, or should be, as important to the photogrammetrist as is a knowledge of their uses. For engineering purposes spot heights might be required with a probable error of $\frac{1}{10}$ m. This can be and has been done, but requires photography under certain conditions (regardless of the apparatus available), and the photogrammetric engineer should be able to specify what these conditions are.

Again, one might be required to make a topographic map, to a specified contour interval, from existing photographs. We are now in a position to say whether or not this can be done. Table 10 will give the answer if the necessary ground control can be established. If more advanced equipment, for example, the multiplex projector, is available the advantages are reduction in the amount of ground control required, and higher speed of compilation. Accuracies, too, are better with such equipment, so that a greater flight altitude would be specified for the same precision, thus greatly reducing the number of overlaps required for a given area.

LIMITING VALUES OF H. We have seen that the lower the aircraft the better the precision, but there are limits below which the aircraft cannot fly and obtain suitable pairs. There is a physical limitation set by the terrain itself, and, but entirely apart from this, there is the minimum exposure interval and the aberration effect of the motion of the aircraft while the shutter is open.* Both of these limit the minimum height.

It may be shown that H min. = 3·67 sif/l, where s is the speed of the aircraft in miles per hour, i the least exposure interval to which the camera may be set, f the focal length in inches, and l the size of the format in inches, measured in the direction of flight.

Exercise 25. Find the best parallax bar heighting precision to be obtained from a 12 in. 9 × 9 cone, speed of aircraft 200 m.p.h., minimum shutter interval 3 sec.

$$H \text{ min.} = 3 \cdot 67 \text{ sif/l} = 3 \cdot 67 \times 200 \times 3 \times \tfrac{12}{9} = 3000 \text{ ft. nearly.}$$

Table 10 indicates a probable error of about 2 ft., under good conditions. This assumes that the ground itself may be seen, and that adequate vertical control is provided.

MOTION ABERRATION. Motion aberration results from the movement of the aircraft during the short time the shutter is open; vibration is negligible if the camera be well mounted but otherwise can cause serious trouble. If the photo air-base is 90 mm., and the shutter interval 3 sec., the speed of aircraft, at photo-scale, is $\tfrac{90}{3} = 30$ mm./sec. If the exposure is $\tfrac{1}{300}$ sec., the image movement is thus one-*tenth* of a mm., and we are trying to measure parallax to $\tfrac{1}{100}$ mm. This motion is in the x direction, the direction of flight.

The effect is rather remarkable, for the fact remains that we can measure parallaxes to the limits given above. It will, however, now be obvious why we do not consider the third figure of a parallax mean as being significant.

Without going further into this field of research, it may be stated briefly that in the case of features like trees the effect is an apparent loss of height. This has been investigated by Andrews(11). Upon coarser features, say the top of a house or the top of a hill, the effect is very much reduced. The problem of really sharp definition in a photograph taken from a low-flying aircraft moving at 200 or 300 ft./sec. (much faster than this in war if the aircraft is to return to its base) is a complex one, linked up as it is with definition and distortion, film speed and grain, and a number of other factors. $\tfrac{1}{100}$ mm. is a small quantity, and the problem of obtaining photographic measurements to these limits, by any means, is a difficult one. The wonder is, not that photogrammetric techniques give results that may be a few feet in error, but that the errors are so very small when one considers all the factors involved.

SONNE MOVING-FILM CAMERA. Loss of definition resulting from motion of the aircraft can be overcome by a moving film camera. Although

---

* Civil practice indicates it is desirable to limit speed of aircraft to H/30 miles per hour with shutter speed 1/200 sec. See also Table 14, page 147. In exercise 25 above, this criterion indicates shutter speed of 1/400.

its photogrammetric applications are restricted, the Sonne camera can produce sharp photography at a very large scale from fast hedge-hopping aircraft. In this camera the film moves at a controllable speed across a fixed focal plane slit. Synchronization results when speed of film is to speed of aircraft as f/H, and under these conditions there is no motion of the image *relative to the film* during exposure. Synchronization may be manual or automatic (compare the Sonne camera, in principle, with a moving-lens panoramic camera). Analogous to an ordinary focal plane shutter, exposure time equals slit width divided by film speed.

Exercise 26. Height of aircraft 150 ft., f 6 in., speed of aircraft 350 m.p.h. Calculate the synchronous film speed in a Sonne camera:

$$350 \text{ m.p.h.} = 350 \times \tfrac{88}{60} = 513 \text{ ft./sec.}$$

$$\text{Scale} = f/H = 1 : 300.$$

$$\text{Film speed } \tfrac{513}{300} = 1 \cdot 71 \text{ ft./sec.}$$

Exercise 27. Same data; what slit width is required for exposure of $\tfrac{1}{300}$ sec.?

$$\text{Width} = (\tfrac{1}{300}) \times 1 \cdot 71 \times 12 \text{ in.}$$

$$= \tfrac{1}{16} \text{ in. (about).}$$

STEREOSCOPIC STRIP PHOTOGRAPHY. By dividing the cone and using two lenses, one covering each (longitudinal) half of the film, continuous stereoscopic strips are obtained if the one lens be positioned in front of and the other behind the slit. The photo air-base is the distance apart of the lenses, and the actual air-base the distance travelled by the aircraft during the time the film takes to move the distance between the lenses. Viewing is by a special stereoscope. Photogrammetric techniques have been developed, and excellent results have been obtained in the measurement of heights. A full description is given by Kistler in [17].

The Sonne camera is a wartime development of the Chicago Aerial Survey Co. for the U.S. Navy and Army Air Corps.

CHAPTER 7

RADIAL-LINE TRIANGULATIONS, GRAPHICAL AND MECHANICAL

The Radial-line Plot, Base-lining and Point Transference

PLUMB POINT AND ISOCENTRE. In a truly vertical photograph θ of fig. 1 is zero, hence the principal axis pSP is vertical and coincides with the plumb line, n and p are therefore coincident. It has been shown that height displacement is radial from n; in a vertical photograph it is thus radial from p since the plumb and principal point coincide.

It has also been shown that tilt displacement is radial from the isocentre i. This point too is coincident with the principal point of a truly vertical photograph. Whereas the tilt displacement in such a photograph is nil, the initial instantaneous direction of displacement due to tilt must be radial from p.

THE RADIAL-LINE ASSUMPTION. We therefore have the proposition that a vertical photograph is angle true with respect to its principal point. This is rigidly true if the tilt be zero quite regardless of the amount of relief. It is very nearly true for slightly tilted photographs where the relief is not great compared to the flying height. This approximation is known as the radial-line assumption. The magnitude of the error involved is discussed below.

NEEDLE PRICKS AND SCRIBED LINES. A fine needle, held in a pin vice, is used as a scriber and pricker. Scribed lines must be fine and must be straight; a *good* 12 in. steel straight edge should be used. Their visibility may be improved by 'inking in' with chinagraph, after which the excess wax is removed with a soft cloth. Needle pricks must be of the smallest size consistent with subsequent re-identification. Pricked points should be ringed with a rather large chinagraph circle, large so that stereoscopy of the adjacent area is not interfered with. The back of the photograph may be suitably annotated.

Skill in the use of the needle must be developed by anyone wishing to do serious photogrammetric work. Pricks should be of such size that a small luminous point is seen when the picture is held up to a light source.

The tendency is to make needle pricks which, to scale, are of astounding size. A good object lesson consists in measuring some rejected work with a low-power microscope (Leitz glass). On a 1:60,000 photograph a needle prick $\frac{1}{4}$ mm. in diameter represents a 50 ft. crater on the ground.

CHECK ON ACCURACY OF BASE-LINING. The base-lining drill has been given on p. 57, and the procedure of p. 81 may be used to check the base-lines as drawn. When the photographs are set in the stereoscope at 90° to the normal position, the topographic detail will not appear in relief. If the base-lines are not on identical detail, the line will appear above or below the ground, perhaps sloping. If H is 20,000 ft. and b 90 mm., 1 mm. of parallax corresponds to 220 ft. If then the line should seem to be about 20 ft. above the ground, as judged by adjacent trees, houses and the like, it means that there is a difference of $\frac{1}{10}$ mm. in the base-lines, and such a difference will easily be detected. This represents good base-lining and should be accepted.

MINOR CONTROL POINTS AND STEREOSCOPIC CHECK. On each photograph pick two points of detail, the co-ordinates of which should be about $x = 0$, $y = b$, and $x = 0$, $y = -b$. Such points are called minor control points, or pass points. The points of detail chosen should be minute, and by minute is meant that the feature is obliterated by a needle prick—such points usually exist; one may mistake them for processing flaws until it is seen that they appear in their proper positions on adjacent overlaps.

These points are transferred, by eye, to the adjacent overlaps, and a small glass may be used to advantage. To check the transference, place the pair on a light table and examine with a stereoscope. The fused point of light resulting from the needle pricks must appear to lie on the ground. Here again an error of $\frac{1}{10}$ mm. should quite easily be seen. One should check y error by rotating through 90°. All points of ground control should now be identified upon the photograph and transferred to the others in which they appear.

The geometric strength of the figure—this depends upon the points being positioned as indicated—is of less importance than is correct transference. The co-ordinates given should therefore be considered only as indicating an area in which a suitable point is to be chosen. Some authorities insist that the whole of the transference operation be done entirely in the stereoscope, but the procedure given will be found to be satisfactory, and to be more rapid. The work should, however, be checked stereoscopically.

A short ray is next scribed through each m.c.p. (minor control point) accurately radial to the principal point. All of the above work must be done with the greatest care. The time spent is but a small proportion of the total man-hours involved in a mapping operation, and carelessness at this stage will result in much trouble later. The aim should be to have accuracy of $\frac{1}{10}$, or at the worst $\frac{2}{10}$ mm. It does require careful work, but once it is understood that $\frac{2}{10}$ mm. is a large error, and that greater errors cannot be tolerated, it will be found that progress is rapid.

Drawing the Radial Line Plot

THE KODATRACE PLOT. Fig. 56 shows a radial-line plot in various stages. The drill is as follows:

(i) Base-line the photos, mark and ray the m.c.p.'s and trigs. Draw a tail to each base to assist in alinement.

(ii) Lay out the photographs in a rough mosaic, and cut a sheet of Kodatrace of sufficient size to cover the strip. If the plot is to be at a scale differing from mean photo-scale, allowance must be made for this. With the Kodatrace over the mosaic, mark the direction of the base-line of photo 1; this is to ensure that the plot will not run off the trace.

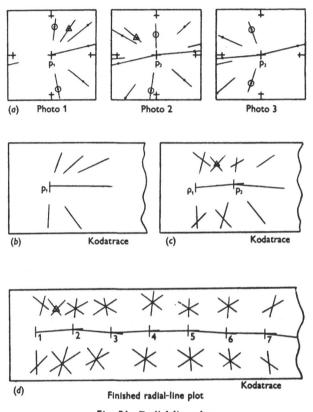

Fig. 56. Radial-line plot.

(iii) Lay the Kodatrace on photo 1, the base-line coincident with the photo base-line, and scribe rays to each m.c.p. and to any trigs. These lines must be scribed with care, and be accurately radial. Mark p_1 on the base-line. (b) of fig. 56 shows this stage completed.

(iv) Lay the trace on photo 2, the base-line coincident with the photo base-line p_2–p_1. Mark p_2, and to each m.c.p. and to the trig. The trace appears as at (c), fig. 48. The m.c.p.'s above p_1 and p_2 have now been located at the intersection of the rays just drawn and those drawn in (iii) above.

(v) Lay the trace on photo 3, the base-line p_2p_3 on the trace coincident with the photo base p_3–p_2. Carefully maintaining this coincidence, slide the trace until the back rays cut the previously located positions of the m.c.p.'s above and below p_2. Mark p_3 on the base-line.

(vi) In this position draw the back rays, which should pass *precisely* through the cuts obtained in (iv). Ray to the remaining points and draw the new base-line, p_3-p_4.

(vii) Continue with the remaining photos. The appearance of the finished strip is shown at (*d*), fig. 56.

SCALE SETTING. It will be apparent that the above plot is a graphical triangulation. In step (iv), when p_2 was marked the scale of the plot was fixed, the distance p_1p_2 having established the scale. In most cases the grid is plotted at or near mean photo-scale, although there are cases where a strip may be run at scale quite different to that of the photograph. To run at a particular scale, determine the distance p_1p_2 at the required scale from the best information available at this time (usually altimeter height only) and set p_1p_2 (fig. 56 (*c*)) to this distance in step (iii). As subsequent adjustment has, in any case, to be made it is seldom advantageous to set scale closely, even when the information is available to do so.

POSITION DIFFERENCES. Working at photo-scale set by p_1p_2, it will usually be found that the trace positions do not coincide with the respective image positions on the photograph. The reasons for this will be evident.

'TILT' ADJUSTMENT. If, in step (vi), the back ray does not pass exactly through the previous cuts, the discrepancy is usually ascribed to tilt—*almost invariably*, however, it will be found to result from inaccuracies of drawing, or of point transference and base-lining. Before proceeding upon any adjustment for tilt, it is necessary to make quite certain that the 'tilt' does not arise from these causes. Having established that the triangle is not due to the errors above, position the trace so that the back rays miss the last intersections by equal amounts and proceed as before. In checking the trace examine each intersection with a glass; the intersections should be exact unless the foregoing adjustment has been necessary. One purpose of carrying the double line of m.c.p.'s is thus seen to be to provide a means of detecting inaccuracies, and of correcting them.

CONSTANT TILT. In this connexion the discussion of relative tilt, and want of correspondence, should be recalled. Absolute tilt cannot be detected without ground control. In the same way if a strip were uniformly tilted, for example, and one were to make a plot from a camera set with a list of 15°, no triangles would appear in the strip from this cause. The resultant plot would, however, not be a plot upon a horizontal plane, but upon a plane normal to the principal axis. It is seen that n, p and i for this tilted datum plane are coincident. If one works from i, the plot is thrown into the horizontal plane—but, working from i there will be considerable error, since height displacement is radial from n.

If a single strip from such a plot be set to ground control, scale is correct in the x direction but incorrect in the y direction. M.c.p. positions may be adjusted by simple geometry, where—as in the trimetrogon system—this constant tilt is known.

THE GRID BOARD. The grid, corresponding to the projection upon which the map is to be made, is drawn up on a smooth surface, if possible large enough to contain the whole area to be mapped. The grid scale is normally the nearest even thousand to mean photo-scale or, for multiplex work, to model scale. In very large mapping projects, size of the grid board seems to be set by the space available, photogrammetrically the bigger it is the better. 5-ply or 7-ply boards, spray coated with enamel and screwed to the floor, are satisfactory. The gaps are filled with plastic wood and sanded. Aluminium sheets have been used with success as have large sheets of good linoleum.

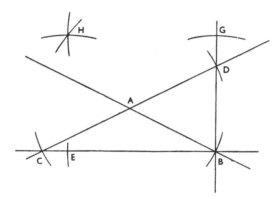

Fig. 57. Master grid construction.

The laying out of a grid is, actually, a small-scale survey operation. A theodolite, good steel tape, and needles to mark the line of sight at intervals, is a satisfactory technique. Where a theodolite is not available, the construction of fig. 57 may be used. On the board draw two diagonals passing through its approximate centre. As the distance will be too long for any ordinary straight edge, these diagonals may be of fine piano wire, or thread, laid down between nails driven in the floor, and alined very carefully so that any irregularities of the board do not introduce inaccuracies. Mark the centre with a needle (slotted templet needles may be used). Mark **AB** such that **B** is suitable as a corner of the finished grid and mark **AC** and **AD** equal to **AB** using a steel tape for these measurements. The angle **CBD** is thus a right angle. Set off the easting **BE** and northing **BG** and, in the case of a rectangular grid, establish the N.W.-corner by measurements from **G** and **E**, checking the measured and calculated values of the grid diagonal **BH**. If the grid is to be geographic, the construction provides axes from which may be set off calculated meridians and parallels.

Upon this grid all horizontal control is plotted.

GRAPHICAL ADJUSTMENT FOR SCALE. A simple graphical triangulation is used to adjust the scale of the minor control plot so that any distance R′S′ on the trace is made to equal RS on the grid. Join RS on the trace and on the grid, and on the trace choose a pole P′ such that R′S′P′ is a well-proportioned triangle. Lay the trace on the grid, R' coincident with R and R'S' alined along RS. Prick through P', and on the grid draw a short-ray radial from this pricked position to R. Repeat with S' of the trace coincident with S of the grid locating P on the grid at the intersection of these two rays.

All the points of the trace are transferred to the grid by a similar triangulation. Three-ray intersections should be used, and a second pole may be advisable where the cuts from the first become poor.

Error and its Adjustment

MAGNITUDE OF HEIGHT DISPLACEMENT. The magnitude of height displacement is calculated as follows (fig. 58). A vertical object ZZ' of height h is photographed from S. The height displacement is zz' denoted by D. The top of the image is at a distance r from p.

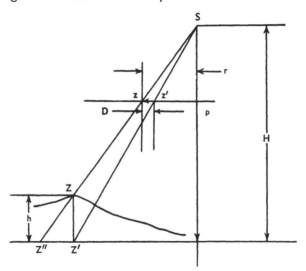

Fig. 58. Magnitude of height displacement in a vertical photograph.

The triangles ZZ'Z" and Spz are similar, hence

$$f/r = h/Z'Z''.$$

But

$$Z'Z'' = DH/f,$$

thus

$$\frac{f}{r} = \frac{h}{DH/f}$$

and

$$D = \frac{rfh}{Hf} = r.h/H.$$

The height displacement for ZZ' is the same as for the hill shown. Note particularly that, as in the case of parallax, the height displacement is independent of f.

Exercise 28. An area of which an air-survey map is required has relief of the order of 300 ft. The relief is greater than this, but is not likely to exceed this figure in a single photograph. Plotting methods are used such that height displacement of more than 2 mm. introduces undesirable errors, format is 9 × 9 in. What should the height of aircraft be, and what precision is to be expected from parallax bar spot heights if there is good vertical control?

On a 9 × 9 in. print, allowing usual overlap, max. r is

$$\sqrt{(1 \cdot 8^2 - 3 \cdot 6^2)} = 4 \text{ in.} = 100 \text{ mm. about.}$$

$$D = r.h/H,$$

$$H = r.h/D = 100 \times \tfrac{300}{2} = 15,000 \text{ ft.}$$

At 15,000 ft. reliability of spot heighting for the conditions stated is plus or minus 10 ft. Higher altitude could be specified, but would result in less accurate heighting. The focal length of the lens specified will depend upon mapping scale desired.

DISPLACEMENTS DUE TO TILT IN COMBINATION WITH RELIEF. Rigorous mathematical analysis is laborious, but the effect may be understood from fig. 59. From S a photograph is taken, the optical axis being tilted as

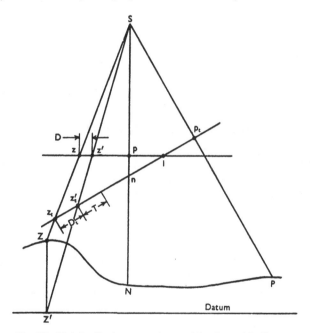

Fig. 59. Height displacement in combination with tilt.

shown. $z_i i p_i$ is the tilted photo-plane. Let a vertical photograph zpi be taken from the same station S. The tilted and the level photo-planes intersect at the isocentre i. An object Z′ on the datum-plane photographs at z_i' and z′ on the tilted and vertical photographs respectively. The tilt displacement is $iz_i' - iz'$, and is marked T on the plane of the tilted picture. If, in addition, there is height displacement, the total displacement is $T + D_t$ as marked. It is seen that the effect is twofold. D on the untilted photograph increases to

7-2

D_t on the tilted photograph as the tilted photo-plane cuts the rays from Z and Z' at greater obliquity and, in addition to this, is the tilt displacement itself. For the conditions drawn, the two displacements are in the same direction—on the other side of i they are in different directions.

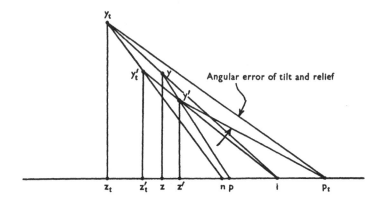

Fig. 60. Superposition of tilted and untilted photographs containing relief, showing magnitude of angular error.

Fig. 60 shows the tilted and untilted photographs superimposed with their isocentres coincident. If z be on a line at right angles to the tilt axis, the displacements are as shown on the line $z_t p_t$. If, however, the point shown at Z' on fig. 59 be at the format position y', the height displacement on the untilted photograph is y'y, radial from p. If now the untilted photograph be tilted, tilt displacement is radial from i and of magnitude yy'_t. The true azimuth from p_t would be measured by raying to y', that is by the ray $p_t y'$. On the tilted view we would ray from p_t to y_t, and the total error due to the combination of tilt and relief is the angle between these rays. y_t is located as shown, radially from n of the tilted photograph. And $y'_t y'$ represents tilt displacement alone.

MAGNITUDE OF ANGULAR ERROR DUE TO TILT AND RELIEF. Referring to fig. 60, a polar diagram of angular error is a distorted figure eight, symmetrical about the principal line $p_t z_t$. The tilt axis is at right angles to this, through i. The maximum of the upper lobe is not along the tilt axis, but (depending upon the magnitude of tilt and relief) at a position where the angle $z_t p_t y_t$ is of the order of 70 or 80°.

In [13] Bagley gives comprehensive tabulations of angular error from which it is seen that with 2° of tilt and 4% relief the maximum angular error is of the order of 9', corresponding to a lineal displacement of but 0·25 mm. at the edge of a 9×9 format. This is the *maximum* displacement—occurring only at two points on the photograph. With 4° tilt the error is about 0·7 mm. —but note as before that error varies from zero at the principal line to reach this maximum only at two picture positions. Thus the probable error of any ray will be much less than the figures given above.

The foregoing, and examination of fig. 60, will justify the common practice of using a substitute principal point. Such a substitute principal point may be a point of well-defined detail reasonably near the principal point—this makes the base-lining operation much more rapid.

ADJUSTMENT OF A BLOCK. It has been shown that the error resulting from tilt, in combination with relief, is small. The reason will now be understood why triangles, ascribed to 'tilt', should be subjected to the most careful scrutiny. Other sources of error are: base-lining, point transference and draughting. The error due to lens distortion is negligible in the methods we are discussing, but paper distortion may cause trouble.

In running a block, it is customary to provide lateral ties by siting the m.c.p. of every fourth or fifth photograph in the common lateral overlap. This necessitates the accurate transference of these particular points, not to two additional pictures but to five. As the lateral overlap may be quite narrow, there is sometimes difficulty in checking such transference stereoscopically.

When a block consisting of a number of strips is laid down, and the common tie points fixed, considerable discrepancies are observed. These result from swing in azimuth and from the scale not remaining constant throughout the strip. These errors arise, not usually from tilt, but more from errors of draughting and from errors in transference and base-lining. Many methods of adjustment have been devised of which one of the most interesting consists of an internal adjustment to bring the various strips mutually into sympathy, followed by an external adjustment fitting the block to the control. This is effected by a series of error contours by means of which the shift of each point, in easting and in northing, may be read off the plot.

CALCULATED TRIANGULATIONS. No such method of adjustment can, however, be considered as being really satisfactory unless the ground control is dense. Since the major portion of the error is from the causes stated, a calculated minor control plot will give better results than are obtainable from a graphical plot, particularly if we have a means of ensuring precision of point transfer and of co-ordinate measurement. Calculations are based upon data from a stereo-comparator from which a precise x and y co-ordinate of each m.c.p. is obtained. The Thomson stereo-comparator is designed for such work; it ensures both precision of point transference and of co-ordinate measurement. A higher order of accuracy is obtainable from such methods than, ordinarily, is required by a mapping scale consistent with the scale of the photography.

The Slotted Templet

THE SLOTTED TEMPLET. The solution to the adjustment problem has been effected by the slotted templet in so far as the mapping problem—contrasted to the application in substitution for third-order survey—is concerned. Its invention is ascribed to Collier. Rather than a Kodatrace plot, templets are used in which are cut radial slots in substitution for rays drawn on the plot. The principal point is a punched hole in which a stud is fitted. One templet is placed upon another and a similar stud fitted at the intersection of the slots, the width of which is a nice fit to the o.d. of the stud.

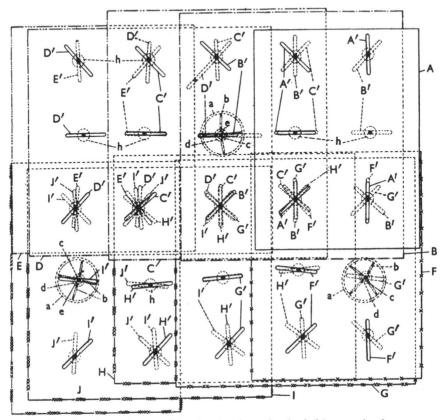

Fig. 61. Slotted templet assembly, showing the spring-loaded intersection locator.

Fig. 61 shows a number of such templets fitted together. Compare this with fig. 56 (*d*). It is seen that it is, in fact, a radial-line plot. Fig. 61 also shows the spring-loaded intersection locator (q.v.).

The assembly may be pulled in and out, corresponding to a change in scale, without altering the angular relationships. The studs are drilled centrally to receive a needle much like an ordinary gramophone needle. A stud having been placed over a grid point representing a trig, a needle is driven in, and the templets laid until another trig is reached. When two trigs are pinned down, the scale is of course set.

Fig. 62 shows the slotted templet cutter as manufactured by Messrs Casella Ltd., London.

Other mechanical triangulators consist of narrow slotted strips with a terminal circular hole. Eight such strips, assembled over the photograph and rigidly clamped at a central boss, form the equivalent of the templets just described. They are used in the same manner. The strips come in different lengths to allow for the long and short picture rays, and longer strips with perhaps two or three alternate central holes permit scale to be changed, or use with pictures of different sizes. They have the advantage that no cutter is required, and that there is no consumption of acetate. They are often used in the plotting of obliques where large sheets of acetate would otherwise be required. On verticals of ordinary size the metal rays may interfere with the intersection of control. The transparent cut templet is preferred for this reason as well as for greater precision.

PREPARATION OF THE TEMPLETS. Preliminary to templet cutting the photographs are base-lined, and minor control points chosen, as in a graphical radial-line plot. The same remarks with regard to careful work apply. Minor control points are, however, always sited on the lateral overlap—thus every picture is tied to all adjoining pictures. Care in transference and checking on the lateral overlap is *essential*. Inaccurate lateral transference is the greatest single cause of trouble in templet work.

Clear acetate 0·4 mm. in thickness (0·016 in.) is suitable, and for 9 × 9 in. prints sheets 10 × 10 in. are convenient. It is better to stock large sheets (say 24 × 54 in.) and to cut to size. The large sheets should be seasoned by hanging.

The drill for templet preparation follows:

(i) Place a piece of acetate on the photograph; weight it down with several lead weights placed clear of the m.c.p.'s. Carefully prick the principal point and any point on each of the short rays drawn on the photograph to each m.c.p. and control point. Ring each prick with chinagraph. *Number the principal point, and mark clearly the direction of flight.* This work may be done on a light table, in which case the photograph need not be rayed.

(ii) Punch the principal point.

(iii) A steel straight edge is supplied with the model illustrated. At one extremity is a hole, to receive a stud, the hole being accurately positioned so that a line scribed with the straight edge will register with its centre. Using this device, place a stud in the straight edge and place the shank of this stud in the punched principal point. The stud will be inverted, and the work should be done on a flat surface (drawing board) in which a $\frac{3}{16}$ in. hole has been drilled to receive the protruding shank. Or the scribing may be done on the table of the cutter.

(iv) Place the scriber in one of the pricks marked (i) above, and rotate the straight edge until it touches the scriber. Scribe the line, and repeat with each point. You should feel the pricked point as the needle passes through it—if you do not, the ray is inaccurate. For plotting at or near photo-scale the scribed line should be about 3 in. long, $1\frac{1}{2}$ in. on either side of the needle prick, assuming that this has been made near the photo position of the image point. Where this straight edge is not used, scribe the lines before punching the central hole.

(v) Place the acetate on the table of the cutter, the cutter stud in the principal point hole.

(vi) Slide the acetate under the blade guard until a circled point is about under the cutter needle, aline the scribe carefully to the cutter needle by means of the sighting telescope, and press the cutter handle firmly down, cutting the slot. Repeat for all points.

(vii) Remove the templet, and examine carefully to ensure that the scribed line (which is somewhat longer than the cutter blade) appears to cut the slot centrally. Reject a templet in which any one slot appears to be in error *in the slightest degree.*

The whole of the above work must be carried out with care and precision, the need for which will be understood after having laid one or two blocks.

ALINEMENT OF CUTTER. If the cutter is not known to be in adjustment, test as follows. Scribe a long line on a narrow piece of acetate and punch a hole near one end, centrally on the scribed line. Place in the cutter, the pin at the end of its run farthest from the blade, and cut a slot. Turn the acetate upside down, carefully work the blade into the slot already cut, and depress the bar to its full extent. This is facilitated if the 'feet' and the telescope are taken off. The scribed line, held in position by the blade, should now cut the exact centre of the sliding pin at both extremities of its run.

Adjustment means are provided, and any discrepancy at either end must be corrected. Remove *half* the apparent error, and recheck. The lock screws may be reached through the slot in the bakelite top; the adjusting screws themselves are on the outside of the frame. Tighten lock screws and outer lock nuts after the adjustment, *and recheck.*

LAYING THE BLOCK. The method of laying the block is self-evident, and no difficulty will be encountered after a little experience. Drive pins at the plotted positions of several adjacent trigs, lay a templet over the first trig and proceed to the next nearest on the flight. As soon as two trigs are included scale is fixed. Proceed in a like manner until the whole area is covered. Any difficulty in the fit of the templets, or any tendency to buckle is an almost certain indication of faulty point transference or an improperly cut templet. When the block, or a considerable portion of it, is finished, it should stand for some time—overnight is usually convenient—after which needles may be driven in the m.c.p. studs, and their positions thus located.

With regard to the accuracy of a slotted templet plot, how well the device solves the problem of utilizing all the various angular measurements to knit them into a triangulation firmly tied to the control will be understood by laying a block better than by any description. Such terms as 'a strong point', 'a weak intersection' and so on will take a very real physical significance, as, indeed, they should. Note that the intersections are of six rays. Just how firmly a six-ray intersection is fixed by these templets is remarkable. It will

Fig. 62. Slotted templet cutter. (By permission of Messrs Casella Ltd., London.)

Fig. 63. A slotted templet assembly of an area of some 5000 sq. km., Northern France. This plot was made in the United Kingdom during German occupancy of France and comprised the control for large-scale military maps. Scale of the master grid is 1 : 30,000.

also be seen that weak and strong rays are nicely given their proper weight in the 'adjustment', that blunders—as distinct from small unsystematic errors—are brought to light, and how a point cannot but take up a position such that the resultant of all the forces acting upon it is zero. In this connexion it is interesting to note that von Gruber[1] states that the method of least squares cannot properly be used in a radial line adjustment owing to the nature of the errors involved. Reference was made to a mathematically calculated plot. It would seem that the templet assembly, by its very nature, effects a logical and accurate adjustment.

But do not give this willing horse the impossible task of effecting an adjustment of the errors of careless workmanship.

Fig. 63 shows a slotted templet plot of an area of about 500 sq. km. in northern France. This comprised the planimetric control for a block of the 1:25,000 series prepared prior to the invasion. Grid scale is 1:30,000, being mean photo-scale.

Calculation of Slotted Templet Accuracy

THEORY OF ERRORS OF A SLOTTED TEMPLET BLOCK. Consider a slotted templet block laid to ground control, and let e be the arithmetic mean error of a number of well-defined image points located by templet intersections, and checked by ground survey. From the theory of errors we might expect e to vary inversely as the square root of the number of control points to which the block is laid—assuming a reasonable distribution of control.

The number of control points remaining unchanged, we would expect e to be larger the greater the number of templets in the block. Experience leads to the belief that it is the lineal distance between controls which is the determinant; thus e may also be expected to vary as the square root of the number of templets. Based upon the foregoing assumptions, we would then have

$$e = k \sqrt{(t/c)},$$

where e is the arithmetic mean error in millimetres at plot scale, k a constant, t the number of templets (or overlaps) covering the area, and c the number of control points. Note that t/c is control density expressed as number of templets per control point. From empirical data the value of the constant k is 0·16—to include a reasonable 'factor of safety'. For e = 0·5 mm., one point is therefore required for each nine templets, for each six if e = 0·4 mm.

Such a formula is of obvious practical value, since it will permit the specification, with exactitude, of the number of control points we must have in a given area in order to map to a specified precision at publication scale. Consider the following typical applications.

Exercise 29. Area 100,000 square miles; required a planimetric map at 1:250,000; specify the ground control.

Consider an aircraft of ceiling 30,000 ft. above the mean level of the ground, and a 6 in. wide angle survey lens. Nett gain per overlap, 23 sq. miles. Number of templets 100,000/23 = 4800 = t. A specification for ordinary good mapping is that mean position error at publication scale shall not exceed 0·5 mm. Hence e = 0·5 × (250,000/60,000) = 2·1 mm., the grid scale being contact scale of 1 : 60,000.

$$c = (0·16/e)^2, \quad t = (0·16/2·1)^2 \times 4800 = 28.$$

Thus 1 point per 3600 sq. miles—points sited 60 miles apart—would be sufficient for the purpose.

Exercise 30.   If stations of a primary triangulation net average 25 miles apart, can good planimetric mapping at 1 : 100,000 be turned out from this control alone?

$$t/c = 25^2/23 = 27·2.$$

Hence
$$e = 0·16 \sqrt{(27·2)} = 0·83 \text{ mm.}$$

The error on the compilation scale of 1 : 60,000 is 0·82 mm., and at publication scale is 0·82 × 60,000/100,000 = 0·5 mm.

Since half a millimetre mean error is tolerable at publication scale, no additional control need be run.

U.S. DEPARTMENT OF AGRICULTURE TESTS.   Extensive templet tests have been made by various authorities. One of the more comprehensive of these is the series by the U.S. Department of Agriculture, and is reported by H. T. Kelsh in (18); an abstract is given in (16).

The Beltsville, Maryland, test area is 155 sq. miles, and in the area were 273 points of known position. Relief 200 ft., photo-scale 1 : 12,000. Prints 9 × 9 in. There were twelve flights of about 19 frames each, and 233 templets were laid. The tests were made with various amounts of control and of control distribution, and in each case the photogrammetric positions of those known points which were not used for control were checked against their survey positions and error measured. Number of control points in the various tests was from a minimum of 4 to a maximum of 31.

Further tests were then made by the same authority, this time in the Bird and Caney watersheds in Kansas and Oklahoma. 4400 sq. miles covered by 2700 templets; compilation (grid) scale 1 : 15,840; from photographs at 1 : 20,000. Test results are in close agreement with the above formula.

Critical examination of such test data led to the development of the error formula.

EFFECT OF POSITION OF CONTROL.   The close agreement between the error calculated from the formula and these tests—in spite of large differences in number and in position of control, and in extent of area—seems to establish the basic principle that, within reason, mean error is not a function of the distribution of the control. That is to say, mean error seems to vary inversely as the root of c whether we distribute the control more or less uniformly, or whether we concentrate it at the edges of the block. Where the control spacing is more or less uniform, the geographic distribution of error may, too, be expected to be uniform. Where it is concentrated at the edges, the larger position errors may be expected to be in the centre of the block. In each case the magnitude of error may be expected to be the same provided there is no undue control concentration.

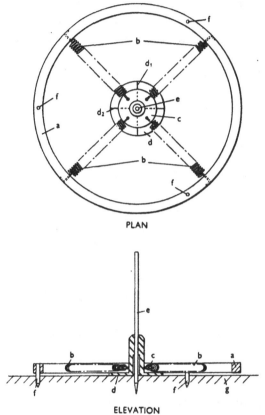

PLAN

ELEVATION
Fig. 64. Spring-loaded intersection locator (floating stud).

The maximum error, and the ratio of maximum to mean, are found to be in accordance with the theory of errors, as need scarcely be stated. Chauvenet's criterion provides a convenient means of estimating maximum error to be expected under given conditions.

CONSTANT k WILL CHANGE UNDER DIFFERENT CONDITIONS. Closer agreement would be obtained with $k = 1 \cdot 5$ or $1 \cdot 4$, but the value $1 \cdot 6$ may be considered as including a factor of safety.

This figure is for use when metal studs are employed. Plastic studs are not (at present) manufactured to the same limits; hence a plastic stud assembly is looser and, accordingly, a large value of k is indicated. There should be no great variation in k over rough terrain, provided tilts are within ordinary limits. With inexperienced operators we would expect k to increase.

ADJUSTMENT TO ASTRONOMIC AND RADAR CONTROL BY THE SPRING-LOADED INTERSECTION LOCATOR. It is sometimes required to map an unsurveyed area from astronomic or radar control which may contain position errors amounting to several millimetres at the grid scale. The slotted templet, with rigidly fixed studs at the control points, can do nothing about this. Unless such control points are widely separated the templets will not lay. When they are forced, they are probably buckled or strained to a greater or less degree, so that there is no real adjustment at all.

The device of fig. 64 is designed for these circumstances. One of these locators, or floating studs, is placed on each control point—fig. 61 shows a slotted templet assembly fitted to a number of locators. The circular frame carries short pins on its under side, and is firmly fixed to the grid board in such a position that the centre of the stud, when free, is at the plotted position of the control. This is facilitated by using a transparent plastic stud with fiducial marks etched upon its base.

The templet is laid in the ordinary way, and when several controls are encompassed the discrepancies in their positions, as plotted and as the templet triangulation would put them, results in a shift of the stud which is free to move under the restraint of the springs. The assembly will thus tend to stabilize with the spring-loaded studs in the position of least total deviation, in accordance with the principle of least work. The plot must be given time to 'creep' before the m.c.p.'s are marked, and the creeping is facilitated if the templets are waxed to reduce friction.

The device gives an approach to a least-square adjustment of the aerial triangulation to the control. It was designed towards the end of the war for use with radar-controlled photography. In a test on the island of Anglesey (Great Britain), where the mean radar error was about 160 m., the error of the spring adjusted templet plot was some 23 m. (19). Flight altitude 27,000 ft., 6 in. metrogon, photo-scale 1 : 54,000.

J. A. Eden reports* modification of this device in such a manner that radar tracking, rather than actual radar fixation, may be used without sensible decrease in accuracy. This greatly reduces the cost of radar control.

* 'Survey Operations with Radar Equipment', Eden, J. A., International Photogrammetric Conference, The Hague, Sept. 1948.

CHAPTER 8

RECTIFICATION MEANS

Rectification

TRUE RECTIFICATION. Air photographs taken with the camera axis inclined to the vertical are defined as tilted. Rectification may consist in a reprojection of the photograph, or negative, in such a manner that image points occupy the same relative positions on the rectification as they would have occupied had the axis been vertical. Thus it is seen that something different to true rectification may be desirable, since true rectification would leave the height displacements as residuals.

OPTICAL RECTIFICATION. In certain photogrammetric work the requirement for planimetry is rather different, namely, that the image points be transformed or 'rectified' to their respective orthogonal projections at the required scale. Although any four points whatsoever may be so projected, a rigidly correct solution is obtainable over an area only when the photograph is of level ground.

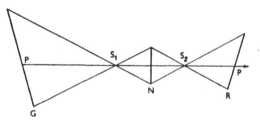

Fig. 65. Projective train, showing rectification.

A projective train for rectification is illustrated in fig. 65, where G represents the ground plane, S_1 the camera lens, N the negative plane, S_2 the projection lens and R the rectified photograph. NS_2R is the rectifier which, of course, is normally quite independent and separate from the camera S_1N.

The Barr and Stroud type ZF-3 epidiascope is an example of a rectifier such as NS_2R (fig. 65). The tilt is shown as the principal plane (of tilt), which requires orientation of the photograph N until its tilt axis is parallel with the tilt axis of R—pS_2 remaining normal to N. Under certain conditions compensation for height displacement can be made by tilting N, or alternatively displacing the photograph from its central position. In the ZF-3 the one tilt axis is on the photo-board, or easel N, and the other on the screen R. To effect rectification on this instrument the following adjustments are required:

  (i) f_1, distance from copyboard to lens.
  (ii) f_2, distance from lens to screen upon which the image is formed.

(iii) X, tilt.

(iv) Y, tilt.

(v) X, tilt of lens ⎫ to satisfy the Schiempflug

(vi) Y, tilt of lens ⎭ condition.

(vii) Orientation of control and photograph.

(viii) Position of control trace and photograph.

Since a change in any one of these settings may require changes in all the others, rectification is laborious and depends quite largely upon the skill of the operator.

AUTOMATIC RECTIFIERS.   Since f_1 and f_2 are mutually dependent, operation of the instrument can be simplified by autofocusing. In addition, the Schiempflug condition must be satisfied, and further simplification results if this is automatic, as in the Carpentier inverter. The inclusion of all these features reduces the number of adjustments from eight to four as under, but at the expense of complexity of construction:

(i) Scale change.

(ii) Tilt about one axis.

(iii) Orientation.

(iv) Position.

Note.   (ii) in combination with (iii) has the effect of reducing the two tilts to a single tilt about the true tilt axis.

There then results fully automatic apparatus such as that of Zeiss (fig. 66). Later instruments by Grant and by Saltzman are now in wide use. These instruments can effect either true rectification or reprojection such that any four points of the rectified photograph correspond to ground control. The scale range from ⅓ to 5x, and tilt to 40° may be rectified. The setting is simple and rapid.

Rectification by Virtual Image

SIMPLE INSTRUMENTS.   If for the real lens image one substitutes a virtual image and uses the eye as the perspective centre, conditions (i), (ii), (v) and (vi) no longer need be satisfied.

The general scheme of such an instrument is as fig. 67; S is the perspective centre, the eye. A 'transparent mirror' (partially surfaced) enables the eye to see the reflexion of the image and, at the same time, the drawing surface. N is a virtual image of the photograph, and the true grid positions of a number of image points have been determined by photogrammetric or other means, as for example C and c.

Scale change is effected by translation of the photograph along the principal axis, and the virtual image is tilted until the photographic control points are in coincidence with their positions on the trace. Note that since N is a virtual image, it may be either above or below the drawing plane G, depending upon whether the compilation scale is larger or smaller than the photo-scale.

Attention is now directed to fig. 68, which is seen to be the equivalent of NS$_2$R of fig. 65, in that the photo-plane is normal to the principal axis. However, in the fig. 67 arrangement, the plane G is held normal to the principal axis, while the tilt is placed upon plane N.

ANHARMONIC RECTIFIER. This instrument is shown in figs. 69 and 70.

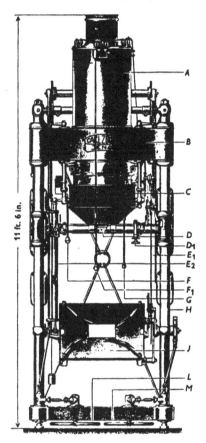

The construction requires no particular comment, except to point out that it is simple and rugged. Collimation consists in setting the 45° mirror at right angles to the principal plane, and in placing the eye vertically above P (fig. 67). When collimated, P of the trace and p of the photograph coincide at all scales and tilts. Through the mirror a reflected and transmitted image are seen simultaneously, and it is empirically found that the reflected image must be sensibly stronger than the transmitted image—the ratio of 5:4 is suitable. A standard commercial desk lamp provides illumination, and the operator, at will, directs the light partly at the photograph and partly on the trace, so that the intensity he desires is easily obtainable.

The optimum working condition is mean photo-scale, and under these circumstances the parallax is small and can be well controlled in that the head-rest enables the operator, without undue fatigue, to keep his head still. Enlargement greater than $\times 1\frac{1}{4}$ involves eyestrain. But the range obtainable is ample to take care of scale variations encountered in practice. Generally, it is preferable to compile at

Fig. 66. Zeiss fully automatic rectifier. (Reproduced from (1) by permission of the publishers, Chapman and Hall Ltd., London.)

mean photo-scale, and to enlarge the compilation photographically (it must usually be photographed anyway) rather than to attempt any great enlargement. Where it is required to transfer detail from a photograph to a map, or compilation trace, at a much larger scale, a suitable lens is placed just in front of the mirror. This case arises in multiplex work where, often, it is more economical to transfer certain detail in this manner. Ordinarily, reduction may be effected without the use of a lens.

OPERATION. It will be noted from fig. 69 that the photo-board tilts about two axes, control of these tilts is by the single knob of the remote control device. As the knob is moved in the direction of the slots of the top slides, motion about one tilt axis results. As it is moved in the direction at right angles to this, motion about the other tilt axis results. Movement in any other direction results in tilts on both axes, in magnitude equal to the X and Y components of the knob motion.

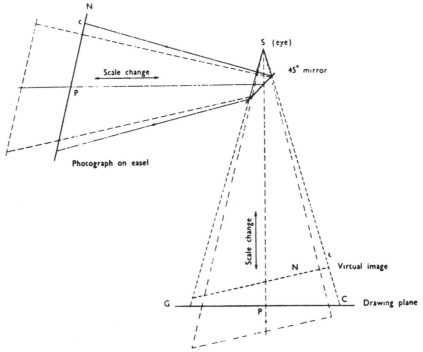

Fig. 67. Diagram of the anharmonic rectifier.

The adjustments are then as follows:

(i) Scale, by sliding motion of the base.
(ii) Tilt, by motion of the control device.
(iii) Orientation of the trace.

The instrument being collimated, the photograph is mounted with its principal point coincident with the intersection of the scribed axis of the Perspex photoboard. No matter what scale is set, nor what tilts are imposed upon the easel, the image of p will remain coincident with the principal point as marked on the collimation sheet.

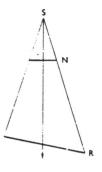

Fig. 68. Alternate rectification arrangement.

A control trace (usually a transparency) is placed on the drawing surface p of the trace coincident with p of the instrument. Approximate scale setting is made while the trace is oriented as well as possible about p.

If the remote control be moved back and forth in the X direction while

Fig. 69. Anharmonic rectifier in use.

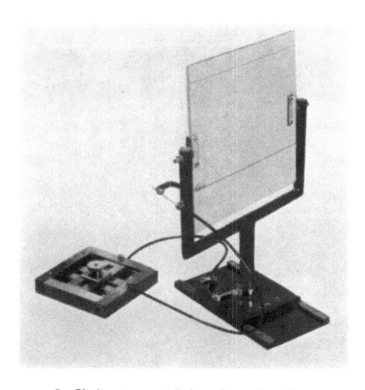

Fig. 70. Remote control device, anharmonic rectifier.

To face p. 112

the Y direction is uniformly changed, all combinations of X and Y tilt are rapidly encompassed. At one setting it will be seen that the control points are more nearly 'on' than at any other. Now a scale change with further slight adjustment of tilt and orientation will usually effect adjustment.

It should be noted that, the instrument having been set, the operator keeps both eyes open. With the right eye he (if right-eyed) sees the reflected image while the left eye, unimpeded, sees the trace. The trace is also seen, though less strongly, with the right eye. This fainter image is, however, sufficiently strong so that the eyes converge, without strain, on the same point of detail. There is no double-image phenomenon.

A 'mirror' image is obtained. This causes no trouble, and is, in fact, in some respects advantageous in map compilation. It would cause trouble were one required to add detail direct to an existing map using the instrument. Under these circumstances, place the print, emulsion side in, on the transparent copy-board, and view by transmitted light. Alternatively, when marking up the photographic detail, place a piece of carbon paper (shiny side up) under the print—or go over the detail with a stylus or fairly hard pencil so that a good carbon impression is obtained.

LIMITS OF THE METHOD. When rectification is used at least four control points—that is to say, the principal point and three others—preferably near the edges of the photograph, are set to control. Thus all other photo-points *in the plane containing the three control points* are homologous with their apparent image positions on the trace. The non-coplaner points exhibit residual displacements the magnitude of which is a function of their distance from that plane.

Where such displacement, calculated from $r.h/H$, is too large to be tolerated, additional points of control must be intersected, and the detail taken off in smaller triangles or quadrilaterals. Note that where difficulty is found in setting to the control points resulting from the radial line plot, if h/H is small the discrepancy is probably due to an inaccuracy or a mistake in the control—unless there are large tilts.

Relief in a single photograph of 3 or 4% is about the limit of this method, for it is uneconomical and unsatisfactory to intersect very many additional control points. Note that h of h/H used in calculating displacements is not the height of the feature above datum, but the height of the feature above (or below) the plane of the three control points. With 4% relief, this distance is usually much less than 4% of H.

PROCEDURE WHEN LIMIT IS EXCEEDED. When these limits are exceeded, neither the rectification method—whether employing an instrument such as the Zeiss, or the simpler instruments—nor the method of direct tracing and 'equating' should be used except for low-grade work.

Under such circumstances recourse should be had to the type of instrument by means of which a true orthogonal projection is obtained. With such apparatus, neither tilt nor relief affect the accuracy of the projection.

Failing this, the photograph may be roughly contoured at suitable intervals—and each contour, together with the detail half-way to each higher and lower contour, taken off with the anharmonic rectifier, each at the correct scale.

It must be clearly understood that no instrument exists which will transform a single perspective into an orthogonal projection.

Other Uses of Rectified Photographs

MOSAICS. When the attempt is made to join a number of photographs coincidence of detail at the join cannot usually be obtained because the detail is not identical on overlapping photographs. If minor control plots are made, and each photograph rectified into these, agreement can be effected unless r.h/H is too great. Again, this is not true rectification, but reprojection to four control points.

PRECISE RECTIFICATION. Precise rectification to the condition of true verticality may be effected by setting calculated tilts on a rectifying camera. As the tilt calculation is a lengthy one, and requires ground control, it seems difficult to justify this procedure except in special cases. The Brock system of air survey, used for accurate large-scale work, uses this general system with specially designed apparatus. Each contour is taken off at a different setting, a precise application of the crude means suggested above. The whole operation, including the aerial photography, is on glass, and the Brock process is capable of work of great accuracy, comparable to that of the better automatic instruments.

CHAPTER 9

SUBDIVISIONS OF AN AIR SURVEY MAPPING OPERATION

Maps and Surveys

AERIAL PHOTOGRAPHY AND AERIAL MAPPING. In common parlance, and, indeed, in technical literature, the term 'air survey' is often used rather loosely when aerial photography is meant. One reads, for example, that an aerial survey of so many hundreds of thousands of square miles was completed. What is meant is that the area in question has been photographed from the air.

Before, however, maps become available there must be ground survey, or the equivalent, the map compilation, and reproduction. The cost of the aerial photography alone is (or should be) of the order of 10–15%, of the total cost of an aerial mapping project, including all the foregoing, a fact which is not very well known.

A thousand square miles can be covered in less than 1 hr. of photographic time (H = 30,000 ft.). Tens of thousands of man-hours of skilled labour may be required for the photogrammetry, mapping, and reproduction. In addition, there may be a season's work for one or more ground-survey parties.

It is well, therefore, to speak of aerial photography, and of aerial mapping, rather than of air survey, so that there may be no confusion as to what, exactly, is meant. Air survey, as used herein, includes the photographic, mapping, and reproduction, operations and, according to the context, the provision of ground control.

DEFINITION OF A MAP. So far as the dictionary is concerned a map is said to be a representation upon a plane surface of a portion, or of the whole, of the surface of the earth. For the purposes of this discussion the word will be restricted to such a representation, but at a scale where conventional signs—which may differ quite appreciably in size from the object represented —are used to depict detail. For example, on the 1 in. Ordnance Survey map of Great Britain, certain roads are drawn as nearly 60 m. wide, whereas in fact they are much narrower. Features adjacent to such a road cannot therefore be shown in their true relative positions. Objects of importance to the purpose of a map may be conventionalized in size. If the map is a good one, such objects will, so far as is possible at the scale, be shown in their correct positions—the centre-line of such a conventionalized road should thus be in its true plan position.

In the same way, a symbol representing a church whose tower is a trogonometric point has a particular part of it registered to the actual position of the point, as nearly as can be drawn and reproduced.

DISTINCTION BETWEEN A MAP AND A PLAN. A plan, on the other hand, is usually of such scale that most objects may be represented in their true size and shape. The British 6 in. map (scale 1:10,560) is then a plan in this sense. The 3906 series, 1:25,000, has been reproduced from a photographic reduction of composites of a number of 6 in. sheets and, in this same sense, too, is a plan. But this is an emergency production. It is difficult to read because too much detail appears, and because many of the objects shown are difficult to recognize owing to their small size. Had it been made for reproduction at 25,000 less detail would have been shown, and there would have been some conventionalization. A photograph is difficult to read, or to interpret, for much the same reason.

Engineering and cadastral plans, with regard to the features they are drawn particularly to depict, show these correctly to scale—and it is for this reason that such scales are large, perhaps 25 in. to the mile (1:2500 or 1 in. equals 200 ft.) or larger. Some important distances or sizes may even, particularly on engineering and construction plans, be dimensioned, and there can thus be no mistake about the size they are intended to be.

Air survey is concerned with the making of maps and plans as above defined.

SURVEY AND MAPPING. Land survey is defined by law, in certain countries, as being the setting out upon the ground of property lines and the like. In these cases the surveyor's plan consists of surveyed lines, their directions and distances marked, drawn to scale it is true, but not with particular precision. A land survey may be looked upon as the drawing of a plan on the ground itself, scale 1:1.

Another kind of survey consists in determining the position of points upon the surface of the earth by geodetic and astronomic methods. Such a survey is the triangulation of England. The primary stations having been fixed to a very high order of accuracy, second-order stations intermediate in position are fixed, usually by trigonometric methods, in relation to the primary stations. These second-order positions are fixed with a high, but less, degree of precision. The secondary net is again broken down, perhaps by traverse, and this forms the basis of control for large-scale mapping.

In this connexion the following dictum of Hotine, as quoted by Hart in (9), is of interest: '...there are those "who consider that geodetic survey and control are a form of useless hair splitting for purely scientific purposes. Actually the ultimate object is no less practical than the prevention of gaps or overlaps in subsequent detail surveys, leading to the same area being mapped twice on different sheets or being omitted altogether."'

In many countries it has happened, unfortunately, that the making of the maps preceded the triangulation. The results of such a practice are remarkable, as anyone will agree who has had the misfortune of being entrusted to effect reconciliation between map sheets made in this manner. Nor is it only the older countries which suffer in this way.

Air survey can enter into a part of this field; for example, the use of the Thompson stereocomparator in substitution for third-order survey has been mentioned. As well there are examples of successful cadastral air surveys.

But by far the largest application of air survey is in the making of maps and plans, preferably between control fixed to the degree of accuracy indicated above, and it is to this application that the techniques described, and about to be described, are adapted.

In military mapping of enemy-occupied territory, air survey is, of course, unique.

Interpretation for Mapping

INTERPRETATION IN RELATION TO SCALE.   By interpretation is meant deciding upon what feature is represented by relevant detail on the photograph. As well as deciding upon what is represented by photographic detail, it is of nearly equal importance to decide what shall not be shown on the map. The photograph shows a wealth of detail; in fact, the whole of the detail that exists on the ground, and not all of this is, in general, required for the purpose of the map. Air-survey maps will usually show more detail than the corresponding ground-survey map, particularly at small scales, and this can be overdone.

Within limits, all well-designed maps should show the same detail density, regardless of scale. If there is more detail per unit of area on a small-scale than on a large-scale map, then either the small-scale map is cluttered up with detail that cannot properly be shown at that scale, or the large-scale map does not show the detail justified by its scale. This rule breaks down at very large scales, for such scales are often specified merely to have enough paper upon which to plan the project.

There is a relation between map scale and desirable scale of the photography; large-scale photographs would not, by choice, be used to compile a small-scale map. A photograph at or near map scale will usually show all the detail required; in fact, in organized areas, it may show a great deal more than can be put on the map.

MARKING THE PHOTOGRAPHS.   Interpretation must be done stereoscopically. Stereoscopes of the type shown in fig. 35 are preferred, and the detail to be included in the map is inked in, on the photograph, using coloured poster inks. The interpreter must be able to recognize the ground features as they appear on the photograph, and must know what to include and what to reject. Photograph reading, like map reading, is not inherently difficult. There is one way to learn both well, get out on the ground.

THE MILITARY PROBLEM.   In the military mapping application the ground being mapped cannot be occupied; the interpreters should therefore have had previous ground training, preferably in country as nearly similar as possible to that being mapped. For example, prior to mapping a particularly difficult area—from the interpretation point of view—of the then enemy-occupied coastal polders of Holland, interpreters previously examined similar areas in our occupation (with similar photographs) so that interpretation of the actual enemy area was facilitated.

The Intelligence application in wartime requires a different kind of skill based upon knowledge of the photographic appearance of enemy defences. Another type of information is based upon examination of enemy lines of communication, dumps and so forth.

THE GROUND CHECK. Photographic interpretation is frequently amplified by examination of the ground, the extent of such examination being a function of the scale of the mapping, the nature of the ground, and kind of map required. Meuser has suggested that this ocular examination could be made with advantage from a helicopter flying at a low altitude.

In some cases information can be obtained from air photographs which cannot be obtained at all on the ground; for example, in certain archaeological work. Again, old river beds and other features of interest to the geologist can be discerned more readily from air photographs than from the ground.

Interpretation for Special Purposes

TOPOGRAPHIC BASIS. Special maps required by the geologist and the forest engineer, and many other maps for scientific or engineering purposes, require to be based upon a topographic map showing faithfully the shape of the ground. When such maps are available beforehand, field work may be greatly reduced.

Accurate topographic shapes, particularly in unexplored territory, are obtainable by photogrammetric methods. Notwithstanding that this is a photogrammetric task, it is done much better by a topographer with a knowledge of how land shapes are formed, and of the significance of such shapes (see (6), Chapter VI, on the relation of geology to topography).

INTERPRETATION OF A SPECIAL NATURE. Interpretation to be included in a special purpose map is the task of a specialist in that branch. The forester will outline on the photographs the boundaries of types of cover in which he is interested, and later this will be amplified by examination of the ground. So with the geologist, the ecologist and so on. An interesting example is given by Chapman in (14), which describes the mapping of Laminaria beds by air survey.

The specialized knowledge of such a scientist based perhaps upon many years of study and experience cannot be expected of a photogrammetrist, but knowledge of simple photogrammetric methods, such as those described in this book, may readily be grafted to his own store. In fact such knowledge might be regarded as a basic part of the training of an engineer or of a scientist, as is a knowledge of English, of physics and of mathematics.

Subdivisions of an Air-Mapping Operation

INTRODUCTION.   We will consider the mapping operation from the point where photographs, or negatives, are available to the finished map in the form of a transparency or transparencies suitable for reproduction by lithographic processes.

The ground-control operation, where the photogrammetry is not based upon existing control, and the specification with regard to the photography, are not discussed at this time.

However, specialized knowledge is required for the management both of the photography and of the ground survey, both of which are, in this case, a means to the end of making a map.

Using simple methods of air survey, most of which have been described, the whole breaks down into the following major subdivisions.

CONTROL (PLANIMETRIC).   The control operation consists in fixing, by photogrammetric means, the number of photo-points required by the particular techniques being employed.

This control is tied to ground-survey data, and may be slotted templet in accordance with the procedure of Chapter 7.

INTERPRETATION.   Interpretation consists in deciding upon what ground feature is represented by the relevant photographic detail. Interpretation is done stereoscopically, and the detail to be included in the map marked, in coloured poster ink, on the print. It is usually supplemented by examination on the ground, particularly in large-scale civilian mapping projects.

TOPOGRAPHY.   Ground control, perhaps by barometer, provides spot heights of sufficient density that additional photogrammetric heights may be calculated by parallax observations. Topography is interpolated from these spot heights, and the drainage pattern, either with or without some simple contour drawing instrument. This may be considered as the simple, non-instrumental, method.

COMPILATION.   In the somewhat restricted sense of the word as used herein, compilation is the preparation of an orthogonal projection, at the scale of the master grid, of the interpreted detail marked upon the photograph. The anharmonic rectifier is used for this operation.

FAIR-DRAWING.   The manuscript compilation is not suitable for reproduction as a finished map. It is enlarged to $1\frac{1}{4}$ or $1\frac{1}{2}$ times the publication scale, and a number of ferro-prussiate impressions taken on enamel paper. From these, a fair-drawing is made, one for each colour which is to appear in the finished map. The topography or the water may be on transparencies for reasons of registration.

PHOTOGRAPHY AND PROVING.   Each colour plate is now photographed on glass, at publication scale. From these negatives (preferably without any retouching) zinc plates are prepared and a number of colour proofs run off.

PROOF-READING AND PHOTO-WRITING.   These proofs are distributed to proof-readers who carefully check the entire area from the point of view of mapping and of reproduction, not excepting the marginal areas of the sheet outside of the map proper.

Meanwhile the photo-writers have been engaged upon the duffing operation, at the completion of which portions of the proofs have been read and they may start on the correction operation itself.

FINAL PROOFS.   The first proof thus corrected is reprinted, and may again be subjected to the proof-reading operation above, and receive such additional photo-writing as is necessary. The final proof having been approved, the negatives are ready for printing.

A Specific Mapping Procedure

CIVIL AND MILITARY MAPS.   If a map made for civil purposes contains inaccuracies or errors (commensurate with its purpose), inconvenience and, perhaps, in the case of a plan for engineering purposes for example, serious financial loss may result. Corresponding errors in a military map will cause inconvenience and financial loss it is true, but there are the graver consequences of ineffective employment of men and of material of war with unnecessary and increased casualties, perhaps to the extent of jeopardizing the whole operation. The responsibility of a maker of military maps is thus seen to be a heavy one. The map tells a story to those who read it, and the report it gives must be a true one.

In the actual making of the military map—reference is made to the making of such maps during, and not before (or after), the occupation of the map area by the enemy—the specific difficulty is that ground examination cannot be made by the map maker or by the surveyor. The military cartographer is thus required to make a better map, and to make it from data less complete. Not only that, but he may be required to make it within limits of time quite unprecedented in civil map-making practice—days as contrasted to seasons or years—and he must do this with personnel who, for the most part, may have had no previous experience either of cartography or of photogrammetry.

MILITARY AIR-SURVEY MAPPING.   The art of air-survey mapping not being very old, it is to be anticipated that there will have been advances along the lines indicated in the previous paragraph during the continuance of World War II. Naturally the problem received attention in the between the wars period, from Hotine in England and Burns in Canada to mention only two, and from many in the United States.

There is described in detail in Appendix 1 the mapping technique developed by Royal Canadian Engineers to fulfil the specific functions mentioned above.

STANDARD MAPPING PROCEDURE (1:25,000) OF APPENDIX 1. The application of this procedure to similar civilian projects is direct. The only real difference results from the possibility, in peacetime, of ground examination before final completion of the map. Just how, and in what stage of the mapping, this examination should be made depends upon the nature and the purpose of the map. It is evident, however, that air-survey mapping is more than the mere employment of air photographs as plane table sheets.

Appendix 1 is an R.C.E. Air Survey Technical Instruction, slightly abridged and modified. It applies to mapping at 1:25,000 from photographs at approximately 1:30,000, f 12 in. format 7×9 in. The detail, shown in black, includes communications, works of man, culture and vegetation. Form-line interval is 10 m. and contours are in the usual brown. Water is shown in blue, and by blue stipple for lakes and for the sea.

Planimetric control was to French trig, upon the values of which a great deal of work had been done by the Geographical Section of the General Staff to bring them into mutual sympathy, and in calculating their Cassini coordinates. In general, the planimetric control was satisfactory; most of the trouble encountered resulted from difficulty in exact identification of the trigs on the photographs.

Vertical control was from spot heights of the hachured French 1:80,000, and in general these spot heights seemed to be strong but were not sufficiently dense for simple photogrammetric methods. Table 10 gives average parallax bar error of 25 ft. or 7·6 m. for this height but with well-conditioned ground control. The contour interval of 10 m. is then not justified by photogrammetric considerations since the error is likely to be greatly in excess of 7·6 m. under the actual control conditions encountered. The 'contours' were thus referred to as form lines, to be considered as showing shape, but not elevation. Where multiplex photography was available, the topographic representation was greatly improved and, at the same time, expedited. Using the multiplex, 10 m. contours were obtained to ordinary civil specifications of accuracy.

CHAPTER 10

PRINCIPLES OF STEREOSCOPIC PLOTTING INSTRUMENTS

Height Displacement

EFFECT UPON SHAPES. It has been shown that height displacement can be calculated from $r.h/H$, and it has been mentioned in connexion with obliques, p. 4, that a change of shape may result, that a surveyed straight line on the ground will only photograph as a straight line on the photograph if, either it is a geometric straight line or, if it passes through the plumb point. The line will not be geometrically straight if there is relief; it is but the trace of a vertical plane with the undulating surface of the ground.

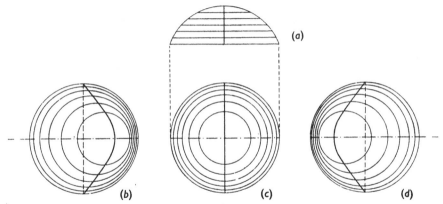

Fig. 71. Showing how relief changes the apparent shape of a feature.
(After Hart, by permission.)

Fig. 71 is reproduced after a diagram given by Hart in (o). At (a) the hemisphere is shown in elevation, with a number of contours and a central meridian. (c) is the true plan view, how the feature should be shown on a map. At (b) is a photograph taken from an aircraft approaching the object at a low altitude, so that h/H is large, (c) is the picture taken from directly overhead, and (d) is taken when the aircraft has passed over.

If we consider this as analogous to a road crossing a knoll, it is seen how its shape may appear concave, straight, or convex, depending upon the position from which it was photographed. Actually (c) is a true plan view and has not been drawn in perspective; the top circles being nearer to the camera would appear larger in an actual photograph. But it well illustrates the principles involved.

SIMPLE OR RECTIFICATION METHOD. In the simple methods we have been considering we have, implicitly, assumed that the 'rectified' photograph shows true shape. To keep the effect within permissible limits two devices were used: first, it was specified that h/H should be small; secondly, we took care to work, as far as possible, in the centre portion of the photograph. For the conditions postulated, the method is satisfactory—in fact, an experienced photogrammetrist, working near the limit of the simple method, will carefully examine the shape of the road as it appears in successive overlaps, on the watch for the effect shown at (*b*), (*c*) and (*d*).

The possibility has been mentioned of raying in a sufficient number of points on the feature to establish its true shape, and of taking off each contour —and the detail lying midway between it and the next higher and lower contour—at a different scale setting in an epidiascope, or in the anharmonic rectifier. The Brock process uses this second method (much refined), and the first method is well known.

An infinite number of intersected points, or an infinite number of scale settings, would give a true orthogonal projection. But it will be realized that a device to give a true orthogonal projection of detail and at the same time to draw contours at their correct position in orthogonal projection, would simplify the plotting. Such instruments are the Zeiss stereoplanigraph, the Wild autograph, the Zeiss multiplex projector, and many others. They perform these operations (substantially) without calculation and regardless of tilt or of relief, effecting this by spacial intersection of an infinite number of rays.

Principles of Automatic Plotting Apparatus

Definition. By an automatic stereoscopic plotting instrument is meant one in which the motion of the floating mark will cause a true orthogonal projection to be drawn as it traces the feature, maintaining apparent coincidence with the stereoscopic model. Contours are drawn in the same manner, the floating mark being set to a height corresponding to the particular contour desired. Both detail and contour are drawn in their correct orthogonal projections without error due either to tilt or to relief.

Geometrical Considerations. Consider a ground point A photographed from each of two air stations S_1 and S_2 (fig. 72). The photographs are not vertical; in fact, they may have any tilt and each camera may be at any height above datum. A is any ground point whatsoever in such a position that it will appear in both photographs, a_1 is its image point on photograph 1, and a_2 on photograph 2.

Neglecting distortions, AS_1a_1 is a straight line as is AS_2a_2; since they intersect at A they must be in the same plane, which plane also contains the line S_1S_2—the air-base. Thus the object A, the perspective centres, and the photographic images of the point are coplanar. And since A is any point the proposition is a general one.

Such a plane is called an epipolar plane, although it is somewhat simpler to call it a (air) basal plane. We have then that any two corresponding negative image points, the two perspective centres, and the ground point itself lie in the same basal plane.

If S_2 is now moved towards S_1 to any intermediate point S_2' without altering relative orientation—that is to say, simple translation only along the base-line S_1S_2—corresponding rays will still intersect, since the rays remain in the same basal plane. Thus the envelope of intersecting pairs in the common overlap will be a replica of the ground, at scale S_1S_2'/S_1S_2.

Plotting instruments of the kind to be described are based upon this rather simple principle; their complexities are optical and mechanical.

GONIOMETERS AND THE PORRO PRINCIPLE. A goniometer may be defined (rather loosely) as an instrument in which a line of sight directed at an image point is made always to pass through the perspective centre. Thus the Canadian High Oblique Plotter is a goniometer, and in this case a telescope is actually pivoted at the position occupied by the inner node of the lens.

Another possibility is to place the negative in the actual camera which took the picture, and to arrange a telescope so that, as it is directed at an image point, the line of sight passes through the inner nodal point of the lens—this is effected by so mounting the telescope that the external line of sight shall always pass through the outer nodal point. In this way, the path of a ray of light from the image point, back through the lens and emerging at the external node, follows the same path as did the ray of light from which the image was formed. Thus distortion in the lens is seen not to affect accuracy of alinement. This very ingenious solution is ascribed to Porro, and is known as the Porro principle.

The original camera and lens are seldom used in devices built upon the Porro principle, but lenses with, as nearly as possible, the same distortion characteristics are employed.

PROJECTORS. If the camera negative were illuminated from the rear an image would be projected but, to bring it to focus at any plane closer than the hyperfocal distance, the internal perspective conditions must be changed. The principle of Porro cannot then be applied in the same manner. The requirement in projectors, as such instruments are called, is that the angle subtended at the external node by any two points of the projected image shall be the same as that subtended, at the external node of the air camera, by the corresponding ground points. This is self-evident, and it follows that narrow-angle photographs may be used in a wide-angle projector without error from that cause—provided only the angular relationship is maintained.

PLOTTING PRINCIPLE. Consider the cameras of fig. 72 replaced by goniometers or projectors, arranged in the same relative positions as before. To S_2 give a motion of pure translation in the direction S_2S_1 to position S_2',

A would move to A' and the scale would now be the instrumental air-base $(S_1 S_2')$ over the actual air-base. As stated before, it is essential that the motion of S_2 must be one of *translation only* along the air-base.

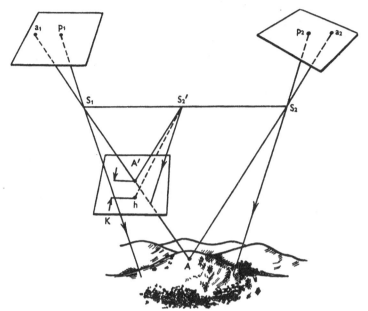

Fig. 72. Biprojective correspondence.

Consider first the goniometer—two High Oblique Plotters if you wish, correctly placed. The plotter space rod is in each case parallel to the telescope. Accordingly, the two space rods will intersect at a point representing the true position, in plan and elevation, of A', as well with an oblique as with a vertical photograph, and quite regardless of the relief. If a pencil were attached to the intersection of the two space rods and so arranged that a point vertically below A' is traced, we would have an automatic plotting machine.

This arrangement would be crude, requiring each telescope to be sighted independently on the same image point. In an instrument on this principle the two telescopes would be coupled mechanically to provide the required motions. The optical arrangement would be such as to present one image to each eye, and thus a stereoscope model would be seen. Floating marks would be included in the optical system and the motion of the floating mark, kept in contact with the model as it traces a feature, would cause the pencil to draw a true plan or orthogonal projection.

To trace a contour, the intersection of the space rods would be restrained in a horizontal plane. Hence as the floating mark is moved maintaining contact with the stereoscopic model, a section with a horizontal plane, or a contour, would be traced, and, moreover, traced in orthogonal projection.

It will be understood that no actual instrument is described or suggested above, but only simplified application of the principles.

Alternatively, use could be made of projectors arranged in their correct relative positions, in which case the intersection of corresponding rays would form a real, not a virtual, image. If one projector were illuminated with blue light and another with its complement red, we could view this real image stereoscopically using spectacles with one red glass and one blue glass. Hence to each eye would be presented only one picture and a stereoscopic model would be seen. Polarized light may be used to separate the images, and it is not clear why it is not in general use. It would seem to have advantages in increased transmission, and in reduction of red fatigue. With colour photography it is the only practicable means of separation.

A screen with a black dot, or cross, will act as a floating mark as it is placed physically above or below the surface of the model. If the screen is supported at a constant height above a horizontal datum plane, a tracing pencil attached will draw a contour as the mark is moved in contact with the model; similarly, the true orthogonal projection of detail may be traced, if the floating mark be kept in contact with the model, raising or lowering the screen or tracing table to effect this. As before, tilt and relief may be of any magnitude.

MECHANICAL AND OPTICAL CONSIDERATIONS. The mechanical means whereby the line of sight is constrained in its proper position are capable of a number of inversions, each of which may result in a different kind of instrument. The mechanical means whereby the rods are actuated, and whereby their motion is caused to move the tracing pencil, differ greatly in various instruments. Whatever the mechanical arrangement used, it must be of the highest precision, and manufacturers vie with one another in working to very close limits indeed. It will be seen that, regardless of how sound such an instrument may be in principle, the accuracy with which the space rods can be made, and the precision with which the proper intersections of the various axes may be effected, and maintained, greatly influence the accuracy of the plotting.

The optical system differs greatly in different instruments, and may be complex and elaborate to provide the necessary adjustments. That of the Wild A-5, for example, contains thirty main optical elements. They, too, must be of high precision and must be capable of maintaining the nicest adjustment. The comparative optical and mechanical simplicity possible with a projector system has certain advantages.

LARGE PROJECTORS. The projection of real stereoscopic images involves a number of problems in connexion with scale, depth of focus, illumination and, as well, of lens design.

Consider the projection of 9×9 in. negatives of a 6 in. lens, or rather the projection of glass contact diapositives. The physical size of the projectors would necessitate their being placed at some reasonable distance apart ($S_1 S_2$, fig. 72) to avoid interference and to allow for tilts, and this in turn necessitates that the image be enlarged to about $4 \times$ contact size in projection.

The projecting lens must be placed with its entrance node in the same position relative to the diapositive as that occupied by the camera lens, namely on the optical axis 6 in. from the focal plane. Accordingly a shorter focal length lens is required to project a sharp image of this size. The following are data for $4\times$ enlargement. S_1S_2', the instrumental air base, 14·4 in., f 4·8 in., projection distance 24 in., model 14·4 × 28·8 in. or thereabout.

Now the Scheimpflug condition is not satisfied, so a considerable depth of focus is essential to allow for relief, and for some adjustment of scale. This necessitates a small aperture and intense even illumination of the field.

To date there can be said to have been no successful economic solution of the problem by means of projectors of this size used in this manner, although the full possibilities may not have been explored.*

SMALLER PROJECTORS. If, however, one makes a diapositive, say 2 × 2 in. from the 9 × 9 in. negative, the focal length of the projecting lens required becomes about 1 in. and the depth of focus is greatly increased. A comparatively simple lens will do the work at a reasonable aperture. This phenomenon will be understood by miniature camera enthusiasts without further explanation—the principle is the same.

On the instrument designed to use 6 in. 9 × 9 photographs the figures are: reduction to about $\frac{1}{4}$, subsequent enlargement to 2·3 times contact scale at the optimum projection distance of 360 mm. (14·2 in.). Usable depth of focus is 270–450 mm., allowing for considerable scale variation. Plotting at optimum projection distance, the distance apart of the projectors is 220 mm. (8½ in.) so that, as the projectors are but 4½ in. in diameter, ample space is available for manipulation. Illumination is by a 100 watt 24 volt projection lamp, with suitable condenser, and the projection lens works at f/16.

The scale of the model is instrumental air-base divided by actual air-base, S_1S_2'/S_1S_2, and is independent of the size of projector.

Orientation

GONIOMETER OR PROJECTOR MOVEMENTS. Consider one of the projectors or goniometers fixed; then to recover the relative orientation which existed at the time of exposure the second must be capable of motion in any direction with respect to the first, that is to say, have all six degrees of freedom.

The x-axis is the base-line; horizontal and at right angles to this is the y-axis, and the z-axis is mutually perpendicular. Motion of translation along each of these axes is provided, and such motions are known as bX, bY and bZ respectively. The perspective centre may also be caused to rotate about each axis—rotation about the x-axis is called tilt, about the y-axis tip, and about the z-axis swing. These axes and motions are shown in fig. 75, as applied to the multiplex projector.

* Since the above was written, H. T. Kelsh has developed a projector type plotter of these general dimensions in which the problems of illumination and depth of focus have effectively been solved. See *Photogrammetric Engineering*, March 1947 and below, Appendix 3.

RELATIVE ORIENTATION. One projector or goniometer may have six degrees of freedom as above, the other remaining fixed. But so long as it is possible to effect these six motions of the one perspective centre with respect to the other, the means is not material. Thus some of the motions may be of one projector, and some of the other. Or both may have all six motions so that one, at will, may use that motion which is most convenient. In addition, the pair must be capable of rotation about two approximately horizontal axes, approximately at right angles, in order to effect absolute orientation or horizontalization. Upon completion of these rotations, *relative* orientation must remain unchanged. As before, this can be done in a number of ways.

Referring now to fig. 72, if relative orientation is maintained as S_2 is moved towards S_1 the two image rays and base remain coplanar. Consider a plane such as h, at A'—the images from both photographs will coincide. If, however, the relative orientation has been disturbed, the images of A will no longer coincide, and it will not be possible to effect coincidence by raising or lowering the plane. The points will appear like this (:); that is to say, want of correspondence (y parallax) will occur. When all image points are in correspondence throughout the whole of the model we have recovered the relative orientation.

A definite drill is laid down to effect this, which drill is applicable to all instruments possessing these six motions.

(*Note.* Having reference to such conditions, *tilt* is now used in its specific sense, above. When a tilted photograph is referred to, tilt is used in its general sense as previously defined, namely, θ of fig. 1.)

DRILL FOR RELATIVE ORIENTATION. The procedure described is substantially that developed by O. von Gruber of Zeiss, and published by him in 1924. It is carried out by removing K at 6 points, conditioned as for a minor control plot. Referring to fig. 73, these points are known by numbers as under: 1 and 2, the two principal points; 3 and 4, the minor control points nearest the operator, below 1 and 2 respectively; and 5 and 6, above 1 and 2 respectively.

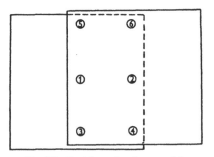

Fig. 73. Position of points used in orientation drill.

Before carrying out the drill for the first time, it is well to study the effect of the various motions in the instrument itself. Where a mark in the shape of at 45° cross is used, K is recognized, as explained on p. 79, by the fact that the two arms of the cross are not coplanar. See also fig. 74; by fusing the left and right pairs of crosses shown, the one arm will appear above or below the other indicating K of opposite sign. Consider the right-hand projector, or goniometer, fixed, then orientation is carried out by removing K at the six points as follows, moving only the left projector.

(i) Point 1, by change in y (bY).

(ii) Point 2, by swing. This is analogous to base-lining, or more properly to setting two contact prints in a ZD-15 stereoscope. Repeat (i) and (ii) until 1 and 2 are quite free of K.

(iii) Point 5, by bZ.

Fig. 74. Stereograms, showing want of correspondence effect.

(iv) Now correct the K remaining point 3 by tilt, continuing until K of equal magnitude but opposite sign appears. That is, overcorrect point 3. The exact amount of correction required varies with the position of point 3, and is given by $(Z/Y)^2$, where Z is its distance below 5, and Y its Y ordinate on the datum plane.

(v) At 1, remove the K now appearing as a result of (iv) by bY as before.

(vi) Repeat (iii), point 5 by bZ; (iv), point 3 by overtilting slightly; (v) point 1 by bY. Continue the repetition until 1, 3 and 5 are quite free of K.

(vii) Check point 2, correct by swing.

(viii) At position 4 with the mark apparently in contact with the ground, remove K by tip which, at the same time, will appear to raise or lower the model as a whole with respect to the mark. Bring it to the ground again by means of bX, or by raising or lowering the floating mark.

(ix) Check point 6. If K appears here, one or more of the points 1 to 5 are not exactly in correspondence and the whole procedure (i) to (viii) must be repeated. It is almost always necessary to go through the orientation several times.

(x) Finally, check all points to make quite sure the orientation is *exact*.

At the conclusion of these operations, all corresponding image points will be in the same basal plane, and the relative positions of the cameras have been recovered.

ABSOLUTE ORIENTATION (HORIZONTALIZATION) AND SETTING TO
SCALE. The model is now a reduced three-dimensional facsimile of that
portion of the earth's surface which was common to both photographs of the
pair—with the exception that dead ground in either view cannot appear
stereoscopically. Its scale is S_1S_2'/S_1S_2. But as yet the scale is unknown and
the model is not horizontal. It may be made horizontal if a minimum of
three points of known elevation are included in the model, and the scale
S_1S_2'/S_1S_2 is determinate if the positions of two ground points are known.

The model is initially so nearly horizontal that scale may be determined
before horizontalization (neglecting the versine error) after which the model
is rotated about two axes—without disturbing relative orientation—until the
three model elevations agree with the ground elevations.

Scale may be changed simply by altering the instrumental base S_1S_2'—it
is usually convenient to work at a definite predetermined scale. Alternatively, in certain instruments, scale is changed by a pantagraph mechanism.

CHAPTER 11

THE MULTIPLEX PROJECTOR

The Instrument and its Ancillary Equipment

GENERAL DESCRIPTION. This description has particular reference to the Williamson-Ross SP-3 equipment, but in general is also applicable to the instruments of Zeiss and of Bausch and Lomb. The instrument consists of a number of projectors, miniature replicas of the aircraft camera, mounted on a bar supported at a suitable height from the plotting table, an accurately plane surface. Each projector has six degrees of freedom of motion.

The tracing table consists of a 4½ in. enamel disk bearing the floating mark. The disk is movable vertically through a range of 100 mm. by means of a thumb wheel engaging a vertical threaded column. A vernier reading to tenths of millimetres measures the height of the horizontal disk above datum. A pencil is provided to record the motion of the plotting table as it traces detail or a contour. The projection lamp operates at 24 V., and means are provided to control the intensity of illumination at each projection in use.

In operation, a blue filter is placed in one projector, and a red filter on the other of the pair—thus a blue image and a red image appear on the screen. Viewing these images with red and blue spectacles, the operator sees a stereoscopic model in the ordinary way. Viewing the model in this manner, as the table is raised or lowered, the mark will be seen to rise or fall with relation to the model.

Setting up consists in removing want of correspondence at the six points, in accordance with the drill given in the preceding chapter. In the previous discussion of want of correspondence the use of a floating cross to detect K was explained. The same principle finds application in the multiplex, but in this case a single cross floats above a real model. However, in the same way, minute differences in y may be detected. The arms of the cross are made to appear coplanar and delicate setting is possible (see fig. 74). If the left and centre crosses be fused, the N.W. arm appears to be above the N.E. arm, while if the centre and right crosses are fused the effect is reversed.

The assembly is shown diagrammatically in fig. 75, and fig. 76 is from a photograph of the instrument.

DIAPOSITIVE PRINTER. The function of the diapositive printer is to make the reduced replica of the original negative. The whole optical train is shown in fig. 77, and it should be noted that the printer may be of any convenient dimensions—the only optical condition is that the angular relationships must be maintained. The broken lines show what is required when a camera of narrower field, that is, longer focal length, is used. Note that a different

Fig. 76. Williamson-Ross SP-3 equipment. (By permission of Messrs Williamson Ltd., London.)

To face p. 130

reduction ratio is required for the printer in order that the angular relationship may be maintained between projecting and taking camera. Provision may be made for this in the reduction printer, or, alternatively, a different reduction printer may be used. The condition is maintained when the equivalent focal length of the diapositive is equal to the distance of the entrance node of the projection lens from the stage plate—where equivalent focal length equals focal length of aircraft camera lens, times reduction ratio.

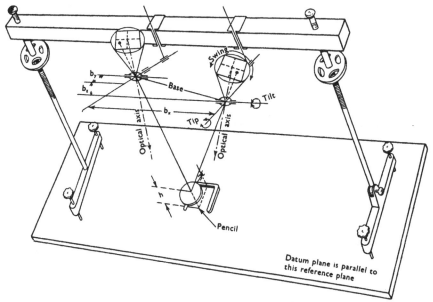

Fig. 75. Diagram of the multiplex projector.

Certain features of the reduction printer are worthy of note. The intensity of illumination at the edge of the field of a wide-angle lens is less than at the centre, and this would cause a corresponding variation in the diapositive density, if uncorrected. Accordingly, an optical wedge is placed in the printer at W (fig. 77), which wedge is light in the central zone and darker at the edges to hold back the under-exposed parts of the negative, the graduation in density being such as to compensate correctly for the marginal effect. The wedge may be moved axially from its normal position, thus providing different degrees of compensation to suit different characteristics and exposures. In the better survey cameras such a wedge is often incorporated in the optical system.

Wide-angle survey lenses have appreciable distortion, and the Porro principle cannot be applied directly. Projector lenses are closely matched for distortion characteristics. A curve may be obtained to show difference in (radial) distortion between the projector lens and the standard curve for the air lens type to be used. This difference is designed into the optics of the reduction printer. From fig. 77 it will be understood that the air camera, reduction printer and projector must all be in sympathy. It is therefore important that the instrument be used with the lens for which it was designed.

A further factor is the variation between lenses of the same nominal focal length. The lens having been calibrated, provision may be made to correct for this in the diapositive printer, so that the equivalent focal length of the diapositive may be in exact sympathy with that of the projector lens. Variations in distortion characteristics of air-camera lenses may cause trouble.

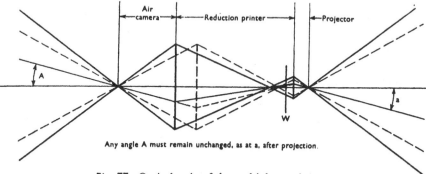

Any angle A must remain unchanged, as at a, after projection.

Fig. 77. Optical train of the multiplex projector.

Air-film emulsions are fast for obvious reasons, and such emulsions are associated with less fine grain. The diapositive, which has practically a standard lantern slide emulsion, is of the finest grain. It can therefore register nearly all that is in the original negative even though its scale is much reduced. Fine-grain techniques of exposure and development are employed, and contrast may be controlled. The result is that, although projected to more than twice the size of the original air-camera negative, the multiplex image is of high quality, provided that the original photography was in all respects suitable.

ELECTRICAL ARRANGEMENTS. For a number of reasons the projection lamps operate on 24 V., and the mains voltage must therefore be reduced by means of a transformer. The transformer output goes to a regulator having four circuits, permitting four projectors to be lighted at one time. The voltage of each circuit may be controlled independently by means of a suitable resistance; this ensures that adjacent images may be of equal brightness even though the diapositive densities differ on the particular common portion being examined.

The 100 W. lamps, naturally, generate heat, and cooling is provided in order that the heat may be carried away other than by the metal and glass parts of the projector where it would have undesirable effects. The cooling arrangement consists of means to pass a strong current of air through the condenser housings.

OPERATION. Orientation drill has been described. When projectors 1 and 2 have been set up, projector 3 may, in the same way, be placed in correspondence with 2, and may be set to the same scale as that of model 1-2 by changing bX until identical readings are obtained on the pass points in both models—and so on throughout the flight. When this has been done, the projectors, and the model, are in the same relative positions as in nature— to a definite scale determinate from comparison to ground control.

The model will, however, not be horizontal, but can be made so if a minimum of three points of known height are identified. These may be two points on the initial model, conditioned as points 5 and 3, and a third near the end of the flight. The model may be levelled in the y direction by means of the foot screws of the bar support until points 1 and 2 show the correct difference in elevation. Levelling in the x direction may now be effected by the bar hand wheels, keeping the readings on points 1 and 2 constant, until the third point is at the correct level.

Other means, to be described, are preferred to effect strip horizontalization, and to set the model scale.

The above outline is, briefly, the original multiplex operational procedure. It should be pointed out that model 1-2 is, itself, usually horizontalized. Horizontalization may be effected by projector movements alone—since each has six degrees of freedom. Practice is to effect approximate horizontalization by projector movements, and fine levelling by the bar motions.

THE SP-3 MULTIPLEX MANUAL. For a description in detail of the instrument, and its care, maintenance and adjustment, reference should be made to the manual supplied with the instrument in use. The remainder of this chapter should be considered as supplementary to such a manual.

Adjustment of Closure in Bridging

EXTENSION AND BRIDGING DEFINED. With dense ground control the model may be set up pair by pair—there would be no necessity to set a number of projectors in mutual correspondence. To set up a single model, two points are necessary to determine scale—and three points of known elevation is the minimum to establish the horizontal plane. Points conditioned as 5, 3 and 2 would be suitable for horizontalization.

Commencing with a strong model set up as above, the strip may be extended for a number of overlaps, and if there is no end tie the strip is said to be a cantilever. When control exists at the end of the strip, the operation is called bridging and the whole assembly could be adjusted to the terminal control. Scale can be adjusted by extending or reducing the overall length, like an accordion, and the several projectors may bodily be horizontalized using the bar as has been described. Except for very short flights, both planimetric and vertical adjustment of bridging should be done as under.

PLANIMETRIC ADJUSTMENT. It has been mentioned that errors of the radial line and slotted templet methods arise from inaccuracies of point transference, of base-lining and, to a limited extent, from tilt. In making a multiplex extension or bridge, minor control points (pass points) are used and serve much the same purpose. If these points, and the principal points, are plotted during the extension, and the plot used to make a slotted templet block, all these errors reduce drastically. Base-lining is automatically a part of orientation, and this is done with greater precision than is possible by other methods. The multiplex plot will, in fact, appear much as fig. 56 (*d*), and templets can be cut from this plot rather than from the photograph or the projected image. In this way point transference errors cannot, it will be seen, occur, since the one plotted point serves the three templets upon which rays to it are drawn. Should the photographs be tilted tilt does not matter—we are no longer working on a perspective projection, but on an orthogonal projection, every point of which is 'angle true'. Tilt of the aircraft camera will thus not affect the accuracy of a slotted templet triangulation based upon a multiplex plot. A second-order error remains, since the unadjusted extension is not projected upon a truly horizontal plane—in extreme cases this error may amount to 3 parts in 10,000, which is quite negligible. The slotted templet plot is also made at twice contact scale; thus the adjustment of such a slotted templet assembly to control will be strong.

Where a number of flights are to be adjusted together templets are cut for each flight, and the whole block laid to the control in the ordinary way. The minor control points* must be identical in overlapping flights—and this source of error, inaccurate transference to the lateral overlap, remains. But identification is easier working on the larger stereoscopic model, and the error becomes less important than formerly.

The m.c.p.'s should be chosen beforehand and marked on the contact prints. They should be marked on one photograph only, which is not only simpler but is less likely to cause trouble. As before, the consideration is positive and easy identification rather than geometric strength of figure.

The above operation results in planimetric positions for six points per overlap—plus the ground control—all in sympathy and all adjusted to scale. It will be understood that the grid upon which the above block is laid will be at or near optimum multiplex scale. The templet adjusts each flight to scale, maintaining at the same time mutual sympathy on the overlap.

The foregoing description is of the method as developed by the 66oth (Topo) Btn. U.S.C.E. The Canadian system uses a slotted templet plot at contact scale. This is more rapid, but less nearly precise.

VERTICAL ADJUSTMENT. Vertical adjustment, or horizontalization, of a strip by means of the bar assumes that the error propagation is a straight line, a common survey assumption, but one which is rather unlikely to be true. It follows from a consideration of the nature of accidental errors that some curvilineal correction will supply a better adjustment in most cases. In the

---

* A 'pass point' is a point common to two adjacent multiplex or other models. Identical height readings on each model maintain constant scale passing from one model to the next. In cutting templets to a multiplex strip as described, pass points on the lateral overlap function as minor control points.

particular application we are considering, it has long been known that where bar adjustment is made of a long flight there are residual errors culminating near the centre of the flight, and experiment on long extensions indicate circular or parabolic departure from the horizontal. In part this is due to the curvature of the earth, and in part to more obscure causes.

It has been found experimentally that an equation of the form $t = kx^n$ fits the actual data, and this provides a means of adjustment in the x or tip direction. In the above equation t is the tip correction to be applied to any point at a distance x from the initial control, k and n are constants.

SEPARATION OF TILT AND TIP.   Before applying any adjustment it will be necessary to determine the error of closure at the terminal model. This closing error results from tilt and tip of the datum plane, and the two effects may be separated as under.

(i) Working an actual example, the terminal model contains two spot heights, A and B, on opposite sides of the base-line. The observed and true values are:

	A	B
True	41·6	40·8
Observed	44·7	43·1

The above are expressed in millimetres on the model.

(ii) On squared paper plot a cross-section on the line joining A and B of the multiplex trace (fig. 78). The section AB is not necessarily at right angles to the base-line. Horizontal scale is that of the multiplex plot, so that

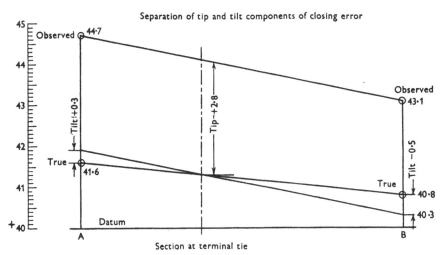

Fig. 78. Graphical method of separating tilt and tip.

distances may be transferred by dividers. Vertical scale 1 cm. equals 1 mm., or 1 in. equals 1 mm. On the section plot true and observed values of A and B as shown.

(iii) Join the points representing true values, and similarly join the plotted positions of the observed values.

(iv) Let c be the centre-line cut of the line joining true values. Through c draw a line parallel to the line joining the observed values.

(v) It is seen that the error due to tip of the model is 2·8 mm., namely, error at c where tilt error is zero, and that the tilt error is plus 0·3 on the left and minus 0·5 on the right. Tilt error is zero at the centre-line.

The graphical solution above is simpler and more rapid than calculation and avoids confusion in determining the sign of the correction.

CALCULATION OF TIP CORRECTION CURVE. In the usual bridging operation there will be control to determine the tip correction at two positions on the base-line. Given these data the equation $t = kx^n$, and hence tip corrections for all points, may be calculated as under. The datum plane is horizontal at the initial model, since we set it accurately to ground control. As we depart from the origin the plane ceases to be horizontal by an amount which is some power of x. This is not due simply to small initial errors of horizontalization but to curvature and to more complex causes; the datum actually becomes curved.

(i) Data: tip correction on the base-line 1·51 m. from the origin, 2·80 mm.; at 1·31 m., 2·20 mm. We have

$$t = ax^n,$$

$$\log t = n (\log x + \log a).$$

Substituting
$$\log 2·8 = n \log 1·51 + \log a,$$

$$\log 2·2 = n \log 1·30 + \log a,$$

$$0·4472 = 0·1790/n + \log a$$

$$0·3424 = 0·1139/n + \log a, \text{ subtract}$$

$$\overline{0·1048 = 0·0651 \quad n = 1·61}$$

$$\log a = 0·4472 - (0·1790 \times 1·61)$$

$$0·2890$$

$$\overline{0·1582, \quad a = 1·44}$$

Hence $t = 1·44 \times {}^{1·61}$.

(ii) Four-figure logs are suitable for this, and the curve may be calculated for a number of points as under. A 6 in. log-log rule is suitable:

x	$x^{1·61}$	f (x)
0·30	0·74	0·21
0·60	0·43	0·62
0·90	0·84	0·21
1·30	1·53	2·20 } check
1·51	1·94	2·80 }

The equation may readily be solved on logarithmic graph paper.

(iii) This curve may be drawn directly on the multiplex plot and the readings of all points corrected in accordance with their x ordinates, the readings being obtained directly from the curve. Note that no correction is made to the initial model, and that the end of this model is taken as the origin for the curve, and for measurements of x.

CORRECTION FOR TILT.　On a given section at right angles to the base, tilt varies from plus on one side to minus at the other, being zero at the base-line, as fig. 78. The tilt itself is taken as varying linealy with x; this is equivalent to assuming the datum plane twists helically. It is probable that this assumption is not quite true, but there seems little justification for any more elaborate calculation. The helical effect is in addition to the parabolic effect of tip.

In the example given the tilt curve will be a straight line from the origin to 0·50 mm. at 1·51 m. from the origin. This curve is drawn on the plot as well. For the example given the sign of tilt correction is minus above the base-line, plus below. The curve gives the correction at the outside edge of the overlap; points nearer to the base-line have corrections varying directly as y. This may be done mentally. For example, a point 0·75 m. from the origin has a tilt correction −0·25 at the upper edge of the strip. If it is one-third the distance out, correction is −0·08, which would be taken as 0·1; y distances are estimated by eye.

Tilt correction is now made to all intermediate points other than those on the base-line. This work is done on the multiplex plot itself.

SPECIAL CASES OF CORRECTION.　Intermediate on the strip may be a single spot height which does not lie on the base-line. Its tilt correction may be obtained from the tilt correction curve—which might be drawn first. The residual is tip. Having only two points on the terminal model, this single point would be used to calculate the tip curve from its tip correction, which we have just obtained, together with the terminal information. In any case, final adjusted values of all control must agree with the true values.

Where the terminal model does not provide data to obtain tip at two reasonably separated x positions, the index n may be assumed from an adjacent strip, preferably run by the same operator.

Where the terminal model contains only one spot height, tilt cannot be estimated from this information and the means of adjustment is from laterally overlapping flights. This, naturally, is less desirable.

Where a block of strips is to be adjusted, there will be two elevations for the pass points, one from each flight. Large discrepancies should be investigated and, in general, reconciliation may be effected. The smaller discrepancies, up to say, 0·3 or 0·4 mm., should be meaned. Only simple common-sense weighting or adjustment should be attempted. Large discrepancies indicate (the extension having been checked) that more ground control is required. If unobtainable, the contour interval must be increased or the work considered of lower grade.

Lakes and shore lines are excellent aids to horizontalization and full use should be made of them where they occur. When there are many spot heights, calculation of the tip curve should not be attempted—draw a smooth curve through all control points.

CROSS TILT. When identical readings are not obtained on pass points, and when the differences are of nearly equal magnitude but opposite sign, the phenomenon is known as cross tilt. This is quite distinct from tilt in its ordinary sense. In the absence of distortions, however arising, pass points may be made to agree, regardless of relief or anything else.

Consider projectors 1, 2 and 3 of a bridge. In practice cross tilt may be due to incomplete elimination of K either in model 1–2, or 2–3, or in both. On the base-line exact agreement may always be obtained; this is simply a matter of adjustment of bX of projector 3. If the outer pass points do not now agree, one difference being + and the other −, the differences being equal or nearly so, then cross tilt is present.

Having made sure that K is *completely* eliminated in model 1–2 check 2–3 carefully. If a trace of K is found, correct, and take new readings. If, after having performed the foregoing with care, agreement still cannot be obtained the residual is due to tangential distortion somewhere in the system, and/or to differences in distortion characteristics between the projectors.

However, even in such a case, agreement can sometimes be effected by swing of projector 3 (remove resultant K by bY). When this expedient is not effective the best that can be done is a 'best mean fit' of the three readings, base-line and each of the two pass points.

It is to be noted that swing of 3 and change of bY will introduce a planimetric swing. But this will not be of great magnitude and can be picked up by the templets.

It is assumed that the following precautions have been taken: inner orientation and diapositive contact with stage plate checked, steps taken to ensure that diapositives are correct, i.e. that collimation marks were exactly on in the reduction printer, that there is no emulsion creep on diapositive, that the printer was set for the calibrated f of the actual camera used, and corrected for negative shrink, if any.

HOW THE ADJUSTED DATA ARE USED. The tip, tilt and templet adjustment operations will result in six control points, both plan and vertical, for each overlap.

From the grid a trace is prepared for each overlap showing the plan positions of the p.p.'s and m.c.p.'s, including any ground control. To this are added corrected readings, and the pairs are set up individually to this control or, as is very much quicker, an operator can add a projector or projectors to his original pair. The relevant topography and detail are then taken off with assurance that the individual plots will fit, since each is based upon control which is in mutual sympathy both on the fore-and-aft and on the lateral overlaps.

The plots are joined together on the master grid in the manner explained under compilation—Appendix 1.

Even though several projectors may be set up together, there is no attempt to rerun the strip on the new control. Better results are obtained more quickly by this method since the laborious and unsatisfactory task of adjusting the set-up of a whole flight is avoided. It is also worthy of note that it is quite practicable to work two operators at each table.

Comparative Precision of Various Instruments

BASIS OF COMPARISON. In the discussion of the ground triangulation analogy to the photogrammetric problem of determining elevation differences it was stated that the comparison is really that of relative effectiveness in measuring angular differences, comparison being made at the same height air-base ratio.

In this connection scale is to be taken as instrumental air-base to actual air-base ratio. There is no advantage in mechanical scale change of this natural scale by means of pantagraph or similar mechanisms—this is no gain in precision, rather a loss.

MEASUREMENT OF THE ANGLE. In the multiplex the angle is recovered by means of a ray of light. Still considering the survey analogy, this time by comparison to a theodolite, we may liken the multiplex projector to a theodolite with a small circle, but one upon which the graduations are rather precise and which we read with a microscope. This is assuming that the diapositive does, in fact, register nearly all the detail of the original plate.

An instrument using a full-size goniometer measures these angles upon a larger circle, and accordingly it should be possible to direct the line of sight to a point with greater precision than that of the multiplex ray of light. (We are assuming no, or equal, distortion in both systems.) Against this increase is the loss in precision associated with the complex mechanical and optical system used to couple the lines of sight to the tracing pencil.

INSTRUMENTAL AIR-BASE. The instrumental air-base of the multiplex is small, say 200 mm. (8 in.), that of the instrument using contact size diapositives might be 400 mm. or 16 in., as has been explained.

Other things being equal the scale criterion would seem to be this, that the natural scale of the instrument (air-base ratio) should be such that the instrumental error of intersection should be rather larger than the error of the drawing mechanism, so that the latter shall not introduce further appreciable errors.

With respect to the drawing of planimetry the multiplex projector does not fulfil this condition for, although the spacial intersection of image rays may be set to one-tenth of a millimetre or better under favourable conditions, the manually operated tracing pencil cannot trace detail with this precision. However, elevations are read with precision commensurate with the inherent accuracy of the instrument.

COMPARATIVE ACCURACIES. Table 11 following gives comparative accuracies of the better-known instruments. For the continental machines the basic data are as quoted by von Gruber from various sources indicated in (1). To form a basis of comparison, the results have been expressed as a fraction of the flying height—when this is not done, such comparisons may be very misleading. In examining published statements of photogrammetric accuracy, one should bear this in mind. Reports of accuracy in feet or metres without a statement of the height of aircraft are meaningless.

TABLE 11*

Instrument	Mean error, fraction of altitude of air station		
	Position	Spot height	Minimum contour interval
Zeiss stereoplanigraph C/2 1928	H/1700	H/6000	H/1500
Wild-Graf Autograph 1928	H/700	H/4000	H/1000
Hugershoff-Wolf Aerocartograph 1929	H/750	H/3000	H/750
Multiplex projector 1945	H/1000	H/4000	H/1000
Brock system 1950	—	—	H/1500
Simple parallax bar, right angles to base-line	—	H/1500	—

It is to be noted that the mean error is given in the above table—maximum errors several times as great are to be expected. The mean error of height of a contour is about twice the spot height error for all the instruments listed, since the operator cannot maintain exact coincidence of the floating mark with the ground. An ordinary specification is that the average error in height of a contour shall not exceed one-half the contour interval, and the contour interval shown is therefore four times the spot height error. The parallax bar error is listed as a matter of interest and is obtained from $0.06 H/b$, where by is 90 mm., implying 9×9 in. prints under the conditions of Table 10.

The multiplex results shown were obtained on well-controlled pairs with good photography and good operators. The mean heighting error, by comparison with spot heights of the 6-in. Ordnance Survey map, was somewhat less than 0.1 mm., with maximum errors of 0.3 mm. At 360 mm. projection distance this is $H/4000$. For less favourable conditions, or inexperienced operators, the above figures must be increased—for all the instruments. Presumably the European tests referred to were made under similar conditions with skilled operators and with the instruments in good adjustment.

Table 11 refers to prewar instruments. The Zeiss C-5 and the present Wild instruments are definitely capable of precision greater than that given in the table. In a recent test by the French Government of the Poivilliers

* See also Appendix 2.

plotter, precision several times as high as the best figure of the above table was obtained. Spot height accuracy of $H/12,000$ is claimed for this instrument under favourable conditions. But it should be pointed out that the Poivilliers system uses a special camera with glass plates—as in the Brock system—and that the goniometer lenses are carefully matched, for distortion, with the air-camera lens.

Air-survey Costs

INTRODUCTION. The cost of an air survey is dependent upon a number of complex factors which vary so much from project to project that a statement of the specific cost of any one operation is apt to be misleading when applied to other conditions. The effect, in broad general terms, of certain of these factors may, however, be discussed.

FACTORS AFFECTING COST. The cost of flying, per hour of flight, is greater at high altitude because of more expensive aircraft and equipment, and because of higher maintenance cost.

Scale varies inversely as H; accordingly the number of frames to cover a given area varies as $1/H^2$. Thus mapping cost per unit of area varies sensibly as $1/H^2$. The planimetric and heighting precision are usually specified on any particular project, and this fixes the flight altitude to be used with the equipment available.

Unit mapping costs per overlap are so much in excess of unit photographic costs that it is nearly always cheaper to rephotograph, obtaining cover designed for the project, than to map from other than optimum photography—even when it is available free.

The nature of the country has a considerable effect upon cost. Rough mountainous terrain will increase the mapping cost per overlap, due to the greater number of contours to be drawn, and will increase the cost of ground control. In highly organized territory, in urban areas for example, the cost of air-survey mapping will naturally increase. The detail density may be such that photography is required at a lower altitude than that set by considerations of precision.

WIDE-ANGLE LENSES. Focal length and height being constant, the number of photographs varies inversely as the square of the tangent of the lens angle. The extremely wide-angle lenses will yield images less sharp, and will have larger distortions than narrow angle lenses.

Considering long and short focal length lenses of the same angular cover, research by Howlett and others indicates that the angular resolution in minutes varies inversely as the root of f. This means that, to obtain the same precision with a $3\frac{1}{4}$ in. wide angle lens as, say, with a 6 in. lens of the same cover, flight altitude would have to be reduced as the root of the ratio of these focal lengths.

Thus the compilation cost for equal precision varies inversely as the focal length for lenses of the same field.

The foregoing argument implies the use of stereoscopic plotting machinery capable of utilizing 'all that is in the negative'. In other circumstances, for example in techniques using paper prints, 9×9 in. enlargements off $3\frac{1}{4}$ in. negatives could be used in lieu of 6 in. contacts from the same altitude, without serious loss of precision.

PHOTOGRAMMETRIC METHOD. Then there is the difference in cost between different methods of mapping. A serious topographic map at any scale cannot (economically) be made except with the aid of instruments such as those we are discussing. Non-photogrammetric mapping costs are, naturally, independent of the number of photographs used in the compilation and, always assuming a map the detail density of which is not a function of scale, depend—for a given style—upon the sheet area. The amount of ground control required is very much greater where 'simple' methods are used than when automatic instruments are used, particularly vertical control.

The Mapping Role of the Multiplex

SCALE. The scale is the air-base ratio, but as height of aircraft above datum is less variable than air-base and, moreover, is usually known; it is more convenient to base scale calculations upon projection distance. Projection distance, to scale, is height of aircraft, thus

$$\text{Scale reciprocal} = \frac{H \times 304 \cdot 8}{\text{Projection distance}},$$

Also, 1 mm. $= H/360$ ft.,

where H is the height of aircraft in feet above datum. At optimum projection distance then scale reciprocal is $0 \cdot 85$ H. It is more convenient to work in scale reciprocals, denoted by R. In approximate mental appreciations think of R as equal to H in feet; this is exact at $304 \cdot 8$ mm. projection distance.

MULTIPLEX PRECISION AT VARIOUS SCALES. Simple mapping methods, using radial line techniques with incomplete correction of the effects of relief, cannot properly be used for high-grade mapping except where h/H is small. A correct solution to the photogrammetric problem results from the use of three-dimensional plotting instruments regardless of tilt and relief.

The precision obtainable depends upon height of aircraft and, other things being equal, upon nothing else. Using multiplex equipment, a given height of aircraft results in an optimum instrumental scale (see Tables 12 and 13). Height variation of plus or minus 10 or 15% is permissible, since the usable depth of focus is such as to permit scale variation of this amount.

At the instrumental scale, good operators can plot well-defined detail within about two-tenths of a millimetre or, say, one-fiftieth of an inch. There are certain other sources of error—including that of the fair-drawing operation—and an ordinary specification for good mapping is that mean horizontal or plan error shall not exceed four-tenths of a millimetre at publication scale. This corresponds to the planimetric precision shown in columns 5 and 6 of Table 12. Such precision is readily obtainable on the SP-3 at optimum scale.

TABLE 12. *Mapping precision at various scales, SP-3 multiplex equipment*

1	2	3	4	5	6	7	8
Optimum height of aircraft	Publication scale		Sheet 18 × 24 in., approx. area (sq. miles)	Plan precision		Min. VI (ft.)	Spot height (ft.)
	RF	Inches		M.	Ft.		
2,800	1/ 2,400	1 in. to 200 ft.	⅛	1	3	3	1
3,000	1/ 2,500	25 in. to 1 mile	⅞	1	3	3	1
6,000	1/ 5,000	—	2¼	2	6	6	1½
12,500	1/10,560	6 in. to 1 mile	12	4	13	15	4
15,000	1/12,500	—	15	5	16	15	4
19,000	1/15,840	4 in. to 1 mile	25	6	20	20	5
30,000	1/25,000	—	62	10	33	30	8
30,000	1/31,680	2 in. to 1 mile	100	10	33	30	8
30,000	1/50,000	—	250	10	33	30	8
30,000	1/63,360	1 in. to 1 mile	420	10	35	30	8

In order to obtain this planimetric precision one point of horizontal control is required on each alternate model if templets are used, but denser control is more convenient. The precision of the control itself must be commensurate with the map scale. For example, at 25 in., control error should not exceed about a foot. If the control error exceeds this, relative accuracy of the map will not seriously be impaired, but the absolute precision will, naturally, not exceed that of the control. The control points should be sited on the lateral overlap.

Just as height of aircraft determines plan precision, so too does it determine the minimum contour interval which can be shown. Column 7 gives this minimum interval in feet, to the specification that mean contour error shall not exceed one-half the contour interval. But—and this is particularly applicable to the smaller contour intervals—this error does *not* include error which may result from the ground being obscured by vegetation. Spot heights on open ground—for example, a profile along the centre line of a road—may be given with the precision of column 8. Note that column 8 is the average error—50 % of the errors may be expected to exceed the figure given, and 50 % to be less.

To obtain this heighting and contour precision, one point of vertical control is required per photograph. These points must be sited on the lateral overlap. The principles above apply with regard to precision of the vertical control. If the vertical control be inaccurate, or be less dense than indicated above, relative accuracy is not seriously impaired although absolute accuracy is.

It is not yet customary to fly photographic aircraft at altitudes over 30,000 ft. Accordingly the precision in feet at scales smaller than 1/25,000 is the same as that at 1/25,000, since such smaller scale maps would be made by photographic reduction of the 1/25,000 compilation. The precision in millimetres at publication scale would be better than four-tenths of a milli-metre in small-scale maps made in this manner. Thus, under such circum-stances, control density might be reduced.

Publication scale does not bear any fixed relation to optimum instrumental scale. Where publication scale is smaller, position error in millimetres on the published map is less than that at optimum scale in the ratio of the reduction. Often special purpose maps, i.e. forestry, are published at scales greater than optimum instrumental scale—largely to give a bigger sheet of paper upon which to lay out the forestry and engineering data. Under these circumstances error in millimetres is, of course, increased with the enlarge-ment, but this can generally be tolerated in such cases. Regardless of whether publication is at a larger or smaller scale than that of the multiplex plot, fair-drawing is carried out at $1\frac{1}{4}$ to $1\frac{1}{2}$ times publication scale to reduce the effect of errors of draughtsmanship and because of the better appearance of the finished work when reduced.

TOPOGRAPHY. Prior to taking off the contours, the drainage pattern is marked on the contact prints (hand stereo) as described in Appendix 1, 4.102, and the multiplex operator works in a manner analogous to that detailed. The topography is still hung on the drainage pattern. No particular comment is needed except the warning that the skill of the operator in keeping the mark accurately in contact with the model is the factor governing precision of the drawn contour.

In flat country it is difficult to locate the position of a contour, tracing with the table. Under these conditions a number of spot heights should be deter-mined and the contour drawn (on the trace) from this information. The shape may be improved by subsequent stereoscopic examination.

The various multiplex plots are joined together on the master grid as described in Appendix 1, 3.3. No attempt is to be made to effect edge recon-ciliation—discrepancies are investigated after the whole has been assembled.

INTERPRETATION. There are the following general cases: mapping of undeveloped country, associated with the smaller scales; medium multiplex scales, dense detail in developed areas; large-scale plans for special purposes, usually engineering plans.

In the first case above, contours and drainage pattern comprise the major portion of the map information. There may be additional information such as, for example, vegetation and forest cover. The interpretation of vegeta-tion, forest type boundaries, and the like requires specialized scientific knowledge. It is neither necessary nor always advantageous that the multiplex operator possess this. Interpretation is done by the specialist working on contact prints. Using poster ink he marks and labels the specific information to be shown on the finished map—and the multiplex operator includes these boundaries in his manuscript. Or the specialist may transfer it to the plot using an anharmonic rectifier.

It is to be noted that the case above is usually associated with quite considerable relief.

In the second case it is inexpedient and uneconomic to attempt interpretation on the multiplex model itself. It is desirable that larger scale cover of the area be available, usually 12 in. if the multiplex cover be 6 in., and interpretation is done on these prints in the manner described in Appendix 1, p. 149. The multiplex plot shows only a network of main features, of such density as to provide control for the anharmonic rectifier or some similar instrument. It is to be noted that this case will usually not be associated with a great deal of relief and that, therefore, rectifier compilation will be quite satisfactory. It is to be noted too that, taking off the detail network as described, there is much more control provided than the nine points of a slotted templet plot—hence there is an opportunity to compensate for height displacement which, otherwise, might be troublesome.

Alternatively, in work of higher precision, the operator uses the interpreted 12 in. photograph (which is nearly model scale) as a guide to the detail to be included in his compilation.

In the third case, large-scale work, the whole of the compilation is done on the multiplex model. This procedure is followed both when topography is the major part of the compilation, and when the detail is dense as is the case in urban work for, say, town-planning purposes. In the latter case, however, supplementary large-scale photography is advantageous.

SPECIAL PHOTOGRAPHY. Conventional specification is 60% fore-and-aft and some 20% lateral overlap. In addition there will probably be strips flown at right angles to the main flight system.

Mention has been made of the difficulty of adjusting the heights of the pass points in the lateral overlap—the 'common-sense' adjustment is not particularly satisfactory on long flights, while no other system seems justifiable. If, however, we fly with 60% lateral, as well as fore-and-aft, overlap the pass points become principal points and the adjustment is accordingly much stronger, since it is only necessary to adjust the principal points themselves.

The slotted templet plot is sensibly stronger and the system has much to recommend it. Note the possibility of running the strips in two directions mutually at right angles.

The extra flying involved is less than might be thought, since the tie flights are no longer necessary.

Where this double 60% flying is adopted, a single 12 in. camera mounted *vertically* will, if exposed at an interval to give the usual 60% overlap in the direction of flight, have 20% lateral overlap. A 12 in. 9×9 is, of course, referred to. Thus a further advantage results in the simplicity with which the larger scale detail cover may be obtained without additional flying and without 'splits' with the corresponding necessity of special rectifying printers or the equivalent.

In all, this two-camera combination with double 60% overlap is ideal for multiplex mapping. It is used in radar work where topographic mapping is required, see (19).

PHOTOGRAPHIC DATA, 6 IN. WIDE-ANGLE SURVEY LENS. Table 13, having reference to multiplex mapping, gives optimum scale, contour interval and coverage of the 6 in. wide-angle survey lens at various flight altitudes. Table 14 gives corresponding flight data. Both tables are calculated for 60% fore-and-aft and 20% lateral overlap.

TABLE 13. *Multiplex scale and coverage, 6 in. wide-angle survey lens at various altitudes*

Flight altitude (ft.)	VI (ft.)	Optimum scale RF	Nett gain per overlap		Overlaps per	
			Sq. miles	Acres	M. sq. miles	10 M. ac.
30,000	30	1 : 25,000	23·2	—	43	—
25,000	25	1 : 20,000	16·1	—	62	—
20,000	20	1 : 16,000	10·3	6600	97	—
15,000	15	1 : 12,000	5·8	3710	172	—
10,000	10	1 : 8,500	2·6	1650	387	6
8,000	8	1 : 6,400	1·6	1055	605	10
5,000	5	1 : 4,000	0·6	410	—	24
2,000	2	1 : 1,700	—	66	—	150

TABLE 14

Flight data: f 6 in., 9×9 in. format, 60% fore-and-aft, 20% lateral

Flight altitude (ft.)	Distance between flights		Air-base		Shutter interval (sec.)		
	Miles	Chains	Miles	Chains	100 m.p.h.	150 m.p.h.	200 m.p.h.
30,000	6·83	—	3·41	—	123	82	61
25,000	5·69	—	2·84	—	102	68	51
20,000	4·55	—	2·27	—	82	54	41
15,000	3·42	—	1·71	—	61	40	30
10,000	2·28	182	1·14	91	40	26	
8,000	1·82	145	0·91	73	32	20	
5,000	1·14	91	0·57	45	*		
2,000	0·46	37	0·23	18	*		

* At these altitudes fly more slowly, if possible.

STANDARD MAPPING PROCEDURE
ROYAL CANADIAN ENGINEERS

Introduction

There are two characteristics which air survey personnel must have, and must use at all times. These are common sense and lack of imagination. We take the view that it is better to have a particular area of a map blank than to have it wrong. Mark the blank 'gap'. Mistakes in a civilian mapping project may cost somebody money. Mistakes in a military map may jeopardize the success of a whole operation with the gravest consequence in wastage of manpower and material.

Mapping from air photographs is not difficult. Although the metamorphosis from the blank paper to the finished three-colour sheet may seem complex and mysterious, the whole may be broken down into a number of unit operations, each of which is simple, and easy to understand. This is a specific contact scale mapping procedure at 1 : 25,000 (about 2½ in. to 1 mile). The principles, however, have application to mapping at other scales.

Mapping at 25,000, as performed by the Royal Canadian Engineers, is divided into seven major parts. These are horizontal control, interpretation, compilation of the planimetric interpretation into a controlled manuscript which we call the pencil compilation, topography, fair-drawing on pulls of the pencil compilation, photography and proving, and, finally, proof-reading and photo-writing.

Simple photogrammetric apparatus facilitates many of these operations, and for each there is laid down a standard procedure, a drill, setting out in detail how and in what order each sub-operation is to be performed. These drills are based upon a considerable experience of mapping of this kind, but nevertheless are not to be considered as immutable—since all progress involves change.

The operation as a whole is designed to take advantage of the benefits resulting from subdivision of labour: reduction of training period, development of high individual skill, and reductions in both the total hours and the elapsed hours per sheet. Such a system requires internal organization which ensures that all parts of the work are, at all times, in phase. It is merely the application of well-known principles of production engineering to a factory designed to build maps.

Procedure in Detail

1. CONTROL

1.10. MASTER GRID

The grid itself is 3 pieces of 4 × 8 ft. 7-ply, screwed to the floor the joints filled with plastic wood afterwards sanded. The surface is again sanded, then sprayed with white enamel, after fastening to the floor.

Grid construction is by the right-angle-in-a-semicircle drill. The diagonals are set out with fine piano wire or thread, stretched taut. A telescopic alidade or a theodolite may be used to ensure that the wire is straight as it rests on slight irregularities of the surface. Measurements are with a 50 ft. steel tape.

The four corners having been set in this manner, both diagonals should check within 0·01 or 0·02 in. Sheet corners are marked on the piano wire block boundaries by pins, and these are joined North and South, and East and West, with piano wire over the pinholes, again using the alidade to ensure exact alinement of the wire. Actual grid lines may be drawn with an H pencil.

1.11. Grid scale is assumed mean photo scale, 1:28,000. Transparent 1:28,000 scales made by photographic reduction of a specially drawn scale are available for use in plotting trigs. All trigs are plotted.

1.20. MINOR CONTROL PLOT

1.21. Minor control is by slotted templet. The first operation is base-lining. Both p.p.'s shall be located and marked as both are needed for rectification. Substitute principal points may be used.

1.22. Base-lines will be checked twice, first by inspection then stereoscopically.

1.23. Minor control requires to be sited on the common lateral overlap for the slotted templet. The usual procedure is followed in transference of m.c.p.'s. Strength of identification rather than geometric strength of the figure is the criterion.

1.24. Identification will be checked stereoscopically, preferably on the light table, on the lateral as well as on the fore-and-aft overlap. When the spot of light does not appear to lie on the model, transference is inaccurate.

1.30. GROUND CONTROL

1.31. Control is, first, strong trigs. That is, trigs the identification of which one can be reasonably certain. Where these trigs cause buckling, investigate. No trig is to be abandoned before careful investigation. Secondly, doubtful trigs. These will usually be trigs which could not be identified with certainty, for example, when the particular part of the building could not be picked out. These are not more likely than the strong trigs to have large errors, but are likely to have errors of the order of $\frac{1}{2}$–1 mm.

1.32. Trigs which have been used for control are marked on the photo with small concentric circles and the name, number and description pencilled on. Black ink.

1.33. M.c.p.'s and p.p.'s, small circle, black ink. Number p.p.'s.

1.34. A trace containing the information of 1.32 and 1.33 is required for use in the rectifier.

2. INTERPRETATION

2.00 INTRODUCTION

2.01. All interpretation is done stereoscopically, stereos magnifying hand are suitable for detail interpretation. In the few cases where there is an area for which there is no stereoscopic cover, there is no option but to mark up the single photo.

2.02. To adjust the eye base of the hand stereoscope, place on photos turned through 90°. The field will probably appear either to be concave or convex. Change the interocular adjustment until the field appears flat.

2.03. Interpretation, with only a few exceptions, is not difficult. The purpose of the following drill is so to systematize the work and the checking that it is difficult to make a mistake, and easy to find the few mistakes or omissions that are made. Mark up nothing of which you do not feel sure. When you are in doubt about an interpretation, see the officer or N.C.O. in charge interpretation. That is what he is there for.

2.1. COLOUR CODE

The operation of marking up the photographs involves the choice of what is to be included in the finished map, and the rejection of irrelevant detail. It also facilitates the operation itself, showing positively what areas have and have not been examined, and is virtually essential for checking purposes. The use of colour simplifies symbolization, and the particular colours chosen are designed to give good visibility in the rectifier. Poster inks are used.

2.11. Yellow. Railways, roads, bridges, houses.

2.111. Railways are shown as single track, double track, tramline, or tramline along road. Use base map for classification.

2.112. Primary roads solid yellow. Other roads broken yellow line. Tracks and paths yellow dotted line. Road classification is done in compilation stage. In built-up areas and towns roads change to red indicating a double line road on the finished map. Do not worry about whether a road is 'primary' or 'other' at this stage. Primary roads for interpretation purposes join cities, villages. 'Other roads' run to farmhouses, connect lesser primary roads, etc. Tracks and trails cross fields and the like. They are sometimes difficult to identify, and are of secondary importance.

2.113. The crossings of roads and railways are of particular importance. It is essential they be clearly marked. Water bridges are shown where you are sure there is a river or creek.

2.114. Houses and buildings. Larger buildings true shape. Houses, etc., small squares about 2 × 2 mm. Orient to face in proper direction. In towns roads will be red, and large buildings only will be marked.

2.12. Red. Roads in cities, villages, built-up areas only—see 2.112, 2.114.

2.13. White. All water features, including shore lines.

2.14. Black. Works of man along shore lines, i.e. docks, sea walls. Any features not specifically mentioned above.

2.15. Green. Woods features.

2.151. Woods (forest) outline in green, in the area put W.

2.152. Brushwood, scrub. Outline in green, in the area put S.

2.153. Orchards, green outline, symbol O.

2.154. Vineyards, green outline, V.

2.155. Trees along roads.

2.156. Fences and hedges (not differentiated).

2.16. Change of colour. On particularly dark prints, or particularly light prints, or areas, it may be advisable to change the above code.

2.2. INTERPRETATION DRILL

2.21. Mark off area to be interpreted. Both fore-and-aft and lateral overlap will be provided to facilitate joining up in the compilation stages. Mark outline in soft lead pencil. This part of the work will usually be done by or under the direct supervision of an N.C.O.

2.22. Put your name, the date, and time you start work in ink on the back of the photo.

2.23. Mark up railways over whole area. Each time a road crosses the railway take time to make quite sure whether road is over or under railway, or whether it is a level-crossing. Mark such crossings with proper conventional sign. (*Note.* Base-map unreliable with respect to crossings.)

2.231. Classify railways as single, double or light (tramways) using base-map classification.

2.24. Roads. Perhaps the most important single feature of a map of organized territory is the roads. The road on the map must look like the actual road on the ground. Shapes at cross-roads, Y's and T's are particularly important.

You see the shape of a road perspectively when driving (or walking) along it. This exaggerates the curves. The map is a plan, and these same curves are lessened in the plan view—that is why it is difficult to relate bends on a road with the same bends on the map. When roads are poorly drawn it becomes nearly impossible to do so.

Again, even if the road shape be marked with meticulous precision on the photograph, each subsequent step, to the fair-drawing inclusive, tends to smooth out these niceties, the original line loses character.

For the foregoing reasons shapes at Y's, T's and intersections will be exaggerated on the original marking up.

Single bends may be exaggerated by making them sharper, and S bends both by sharpening and by increasing the amplitude. The exaggeration does require skill, but more than that it requires common sense. When the principles above are grasped the interpreter should not experience any serious difficulty.

2.241. Mark roads over whole area. Classify by common sense as main roads, other roads, trails—tracks—paths. Do not classify from base-map at this time. Will be done on pencil compilation. Remember roads change to red in built-up areas.

2.242. Check against base-map for tramlines along roads. These generally cannot be seen in the stereo.

2.243. Tree-lined roads are of tactical importance. Mark with green dots along roadside.

2.3. Water. Mark rivers and lakes in white over whole area. Interpretation of the watershed pattern is a part of topography, only water that you can see or that you are sure is there will be marked with the planimetry. Only main rivers, creeks, lakes are required with the planimetric interpretation. The aim is to show all water which must register to any part of the detail plate.

2.4. Remaining Interpretation

2.41. Working in a small area of the picture, say a few square inches, complete all remaining detail. This will consist of houses and buildings, fence and hedge lines, wood features, miscellaneous features (e.g. airfields, cemeteries, quarries). Work systematically, *completing* each small area before proceeding to the next.

2.412. Hedges and fences. Only difficulty here is getting mixed up with minor roads. That is why we mark the roads first. Hint: A minor road usually goes somewhere.

2.413. Houses and buildings. Individual isolated houses and buildings are very important as landmarks. In built-up areas, in towns and villages, smaller buildings will usually be conventionalized—associated with red roads—and only the larger buildings will be shown individually. Where smaller buildings are shown, use a 2 mm. yellow square, correctly oriented. Where building is larger, mark up true shape.

2.5. CHECKS BY INTERPRETER HIMSELF

2.51. Railways again against base-map for classification.

2.52. Tramlines along roads.

2.53. Check all detail on fore-and-aft overlap against preceding photograph. Interpretation must agree exactly.

2.54. As above checks are made, mark back of photograph in ink with the number of each check, followed by a check mark and your initial. Thus, '2.51 check mark initials' means that you have made a final check of item 2.51 (railways against base-map) and are satisfied.

2.6. RESPONSIBILITY

The principle of thus fixing individual responsibility is that errors may be traced to their source, and steps taken to prevent their recurrence.

The interpretation and marking up is the backbone of the whole job. It must be complete and accurate.

3. COMPILATION DRILL

3.1. TRANSFERENCE OF CONTROL

3.11. Fasten a strip of Kodatrace to the master grid matt side down. Strip is 2 in. larger in all directions than area of flight. Seasoned Kodatrace will be used.

3.12. There will be two or three compilation strips per map sheet. The actual size of each strip will be as directed in view of the particular circumstances of each case.

3.13. Prick through the principal points, minor control, and trigs. Ring the p.p.'s with black chinagraph, the m.c.p.'s with blue and the trigs with a black triangle.

3.14. Remove the Kodatrace and ink in the control as follows, numbering the p.p.'s.

3.141. P.p.'s small black ink circles, number each.

3.142. M.c.p.'s small black circles.

3.143. Trigs black cross 6 × 6 mm. with 4 mm. circle.

3.144. Clean off chinagraph with dry cloth; petrol, etc. will not be used because of shrinkage effect.

3.2 SETTING UP IN RECTIFIER, and transferring detail.

3.21. Assemble the instrument, collimate following drill on collimation sheet. Mark the principal point on the centre line as described.

3.22. Place the control plot on the drawing surface, matt side up, with the relevant principal point in coincidence with the p on the centre line. Usually plotting will be done with the base-lines at right angles to the centre line. Hold trace down with lead weights.

3.23. Place the photo on the board, alining the collimation marks with the axes so that the photo p, or substitute p, coincides with the board p. Check by using transparent mount as a light table. The photo will be placed on the board with the base-lines in the same direction as on the plot.

3.24. Place p of the trace on p of the centre line, rotate the plot about p until the bases are in approximate alinement, and set to what you think is, as near as possible, the overall scale. Both tilts zero during this operation.

3.25. Start with the Y tilt fairly large, and turn through the X range. Reduce Y and repeat the X motion. Continuing, combinations of X and Y tilts will have been covered and it will be noticed that at one particular combination the points are more nearly on than at any other. When this setting is found, adjust the position of the plot to its optimum by rotating slightly about p, and change scale as indicated. Now, a slight tilt change, and further small-scale change, will bring the points in. The trace may now be fastened down with cellulose tape.

Notes:

3.251. It is usually possible to set any four points precisely—when this is done, the remaining points will then be 'on' unless tilt and/or relief is great.

3.252. With photography at 30,000 ft., in the Benson area it will usually be possible to set 9 points since the relief is small. If this is not possible it is due either to height displacement, excessive tilt or inaccuracies of control: usually the latter.

3.253. Where trouble is experienced as above, report to officer or N.C.O. in charge.

3.26. Having set the photograph—and this is quicker to do in ordinary cases than it is to describe—trace off the detail in the same order as given for the marking-up drill, namely:

3.261. Trace roads, railways and water over the whole area. At this time ensure that railways are properly classified, that tramlines along roads are shown, and that road and railway crossings are shown as over, under, or level, as marked on the print. Road shapes, particularly at intersections, must be traced with meticulous accuracy.

3.262. Start in one corner and complete all remaining detail in a small area. Work in an orderly fashion through the whole area to be transferred.

3.27. Continue 3.22 to 3.262 with the remaining photographs of the strip.

3.3. Joining Strip Plots, Checking

3.31. Place one strip in position on the master grid setting accurately to the control, fasten down with cellulose tabs. Place the remaining strips of the sheet independently on the grid and, finally, join the pieces to one another.

3.32. Carefully remove the whole from the grid, and cut through the overlap accurately using a *sharp* knife and steel straight edge.

3.33. Replace on the master grid, and butt joint the assembly with tabs of cellulose tape.

3.34. Scribe the grid on this composite.

Notes:

3.331. When the strip plots 'go' special treatment is required, which will be specified by the officer in charge for each particular case.

3.332. Pay no attention to the overlapping detail in making this composite, each strip is to be set correctly and independently on its control on the master grid.

3.35. Investigate any discrepancies of interpretation or position which show up on the cuts, i.e. the lateral overlap. Make good detail at the join where necessary.

3.4. CHECKING OF COMPILATION

The compilation is to be very thoroughly checked. It is required that the compilation shall show correctly and clearly everything that is to appear on the final black (except spot heights). Checking drill follows.

3.401. Obtain a number of reversed ozalids (matt to emulsion).

3.41. Mark one ozalid 'roads and railways'. On this, check as under, using checking form. All amendments will be in black ink.

3.411. Add road classification from base-map.

3.412. Check all railways, including trams.

3.413. Check all railway classification.

3.414. Check trams along roads.

3.415. Check over, under, level, crossings of roads and railways against library photos. See that symbolization is clear.

3.416. Transfer amendments to pencil compilation on light table, bostitch completed check form to ozalid and pass to control stores to be placed on job file.

3.42. Mark another ozalid 'detail', and check against marked-up photos as under, grid square by grid square. Amendments in black ink. As each square is completely checked, initial and date it.

3.421. All houses and buildings.

3.422. Red roads opened.

3.423. Wood features.

3.424. Trees along roads.

3.425. Fences, hedges, etc.

3.426. Miscellaneous. Quarries, cemeteries, airfields, etc.

3.427. Transfer amendments to Kodatrace on light table.

3.43. On detail ozalid, or on another ozalid if convenient, each edge, N., S., E. and W. shall be checked for continuity of detail against compilation of previously compiled adjoining sheets. As an edge is checked, it shall be marked 'Edge checked', dead and initialled. Check as follows.

3.431. Exact junction of all roads and trails, and railways.

3.432. Continuity of road and railway classification.

3.433. Continuity of classification where road has tramline.

3.434. Exact junction of fence and hedge lines.

3.435. Continuity of wood classification.

3.436. Exact junction of water features.

3.437. Transfer amendments to trace on light table.

3.5. CHURCHES

Present policy re churches is that if the photograph shows a building at or near the position where the base-map shows a church, then we call the building a church—whether or not we can positively identify it as such.

3.51. Using detail ozalid (which should then be headed 'detail and churches') or using another ozalid if more convenient, mark each church down on base-map, using the church symbol drawn freehand in ink, and following the policy of 3.5 above. Do this grid square by grid square.

3.52. Where church is a trig which is to be shown as such, i.e. a trig which we have identified and used successfully, it will be on the ozalid cross and circle. Note that churches which are on the trig list, but upon which we have been unable to locate the exact spot used as the trig, or which did not tie in to control, will be shown as churches, not as trig churches.

3.53. Transfer to compilation on light table, freehand and in black ink.

3.6. PREPARATION OF PENCIL COMPILATION FOR CAMERA

3.61. The preparation consists in cleaning up, and in strengthening the line work if necessary. This should be supervised by a draughtsman experienced in this particular work. A good sharp clean pull makes the job of fair-drawing much easier and quicker. The fair-draughtsman is supposed just to ink in this pull. He should not have to puzzle out what a poorly drawn or indistinct line or symbol means.

3.62. After O.K. on linework by Officer in charge Compilation, pass to officer in charge for work order for camera.

4. TOPOGRAPHY

4.011. It takes years of field experience before a topographer becomes sufficiently skilled to translate ground shapes, as they appear to the eye, into a good topographic map. One of the reasons is that we see perspectively; what we see is very like an oblique taken from the same view-point. On the ground, one has first to visualize where the contour goes, then to visualize the change in shape consequent upon the requirement that this be rectified to an orthogonal projection, finally to draw this at a reduced scale. Preferably topography will be done by means of the multiplex projector, but where photography suitable for use in the multiplex is not available, the following is the procedure.

4.012. In training you were given the topographic exercise of fig. 79 and, paying attention to the drainage pattern, to ridge lines, to road and railway shapes, and to cuts and fills on them, you were able to produce quite a good topographic map. This you checked against the published map of the area—and later against photographs viewed in the stereo.

4.013. It is very much easier to draw the contours when the same information is marked on one of a pair of photographs of the area. Now you have not only the data you had previously, but you can fit the shape of the contour to a three-dimensional model which you can actually see. The only difficulty you are likely to have is in flat country, and you then need many more spot heights if contours are to be drawn with the same position accuracy as on the steeper parts. The ground survey party would have the same difficulty, if that is any consolation.

4.101. The topo photographs given to you will have spot heights marked on them. These are from base-map elevations, from parallax readings, or from other information. The first step is to mark off, on each photograph, the area to be worked. Small fore-and-aft and lateral overlap will be provided. This will be done by or under the supervision of an N.C.O.

4.102. First mark up the drainage over the whole area. When you are certain there is running water, mark in solid white. This main drainage shall be marked first. If you mark drainage on both photos it will appear in relief.

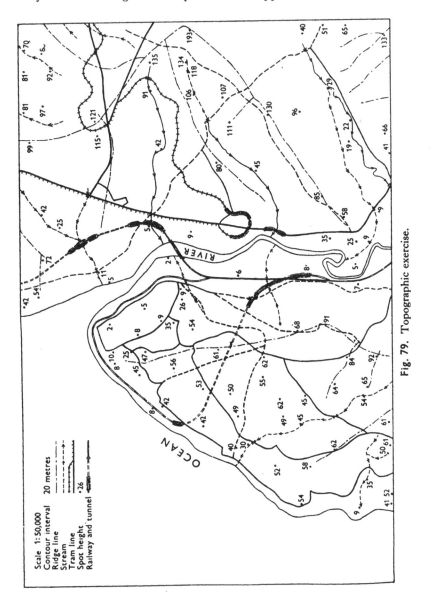

Fig. 79. Topographic exercise.

4.103. Add the minor drainage features in a broken white line. These are valley bottoms which may or may not contain actual streams.

4.104. Railway (and road) cuts and fills are topographic features and will be marked at this time.

4.105. Mark main ridge lines, heights of land, divides, broken yellow line. These, too, may be marked on both photographs, but if this is done care must be exercised that the lines cut identical detail on each.

4.106. Transfer the contour ends from all adjacent photographs. Great care must be exercised at this stage to make certain that all adjacent photos join up correctly. *Number these contour ends.*

4.107. Starting from the contour ends transferred from the adjacent photo, start form-lining an area of reasonable size. The topo prints should be matt surfaced, and a B pencil, kept fairly sharp, should be used. Draw the first contour completely across this area, paying attention to spot height data position, and to the model for shape. This is assisted by the drainage pattern, ridge lines and so on. Complete the area in the same manner. If the country is steep, draw each fifth contour first and fill in the remaining contours; if very steep omit intermediate contours. This work is to be done in the universal stereo.

4.108. Examine the contoured picture (non-stereoscopically) having regard to agreement with spot heights, registration to draws and ridges, road and railway shapes and their cuts and fills. If a road bends, there is probably a reason for it. Does your topography show such a reason, and a reason for the cuts and fills?

4.109. Replace the stereoscope, and check for general rather than for particular agreement with the model, making such alterations as seem necessary.

4.110. Now use a magnifying stereo and, working in areas as before, ink in the topography. The stereoscope must be in adjustment—see 2.02. In the small stereo you will be able to see the photographic detail more readily, and in inking in these smaller shape changes will be followed using the pencil line for general position. The inked topography will thus show more 'character' than did the pencil.

4.111. The contoured overlap is again examined in a general sense as the final check. The following are to be checked in particular, and the check noted on the reverse of the photo together with the date and your initials.

4.112. Contour fit to drainage.

4.113. *Does any water flow uphill?*

4.114. Are railway grades excessive?—for a $1\frac{1}{2}\%$ grade (1:66) 10 m. contours cross at 2·4 cm. distance, scale 1:30,000.

4.115. Are road grades excessive?—contours crossing main roads closer than 4 mm. should be suspected. This represents 8% at 1:30,000 (1:12). Such a grade is not necessarily wrong on a well-travelled road, but make sure it is there.

4.116. The grade-finder, fig. 80, is an aid in the above check.

4.117. Are cuts and fills shown?

4.2. STEREO-COMPARAGRAPHS

4.21. The foregoing system of contouring—form-lining would be a better term—depends upon spot height density to keep the sketched contour level, although it is true that the model helps to some extent.

4.22. A simple stereo-comparagraph (fig. 43) consists of a parallax bar provided with linkage such that, regardless of where it is placed on the format, the centre line of the bar remains parallel to the base-line. To the linkage carrying the bar is attached a pencil. Thus if the vernier is set to a reading corresponding to a particular contour, the floating mark may be moved in apparent contact with the model, and a contour is traced. If magnification is provided, operation 4.110 is automatic.

4.23. Before parallax differences may be taken as representative of height differences there are, as you know, corrections to be made. Where the stereo-comparagraph is used the corrections are worked out (by standard methods) for all the control points, and the photograph is 'contoured' with lines of equal correction. In

GRADE-FINDER
for 10 METRE contour interval

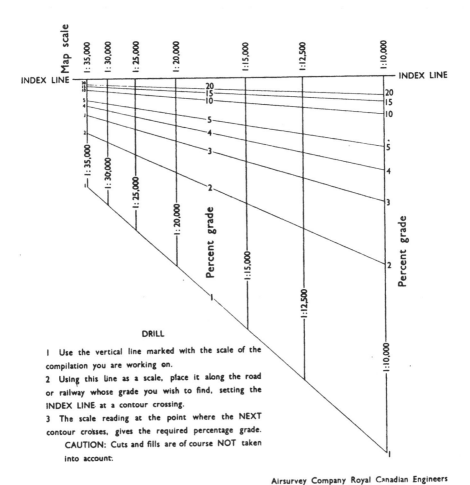

DRILL

1 Use the vertical line marked with the scale of the compilation you are working on.
2 Using this line as a scale, place it along the road or railway whose grade you wish to find, setting the INDEX LINE at a contour crossing.
3 The scale reading at the point where the NEXT contour crosses, gives the required percentage grade.
   CAUTION: Cuts and fills are of course NOT taken into account.

Airsurvey Company Royal Canadian Engineers

Fig. 80. Grade-finder, for use in the checking of topographic compilations. The grade-finder is printed on film and laid over the compilation.

using the instrument the vernier is changed each time and the floating mark crosses a correction contour—or better the correction is made midway between correction contours. This will result in a slight jog, to be smoothed out afterwards.

4.24. Where the control provides a series of pairs of points at right angles to the base-line, intermediate corrections may be worked out by the standard drill at right angles to the base-line. Thompson gives a drill(15), based upon elevations at four control points, whereby correction curves may be worked out for the whole overlap.

4.25. However the corrections are computed, the photographs, once mounted, must not be disturbed until contouring is completed—otherwise the equivalent of further K will be introduced.

4.26. It is to be noted that the trace from such an instrument is still a perspective projection—it is a replica of the motion of the fixed dot of the bar as it follows the detail of the photograph under it. In either method, where the error from this cause cannot be tolerated, the contours may be taken off each at a different scale by means of the anharmonic rectifier.

4.3. TOPOGRAPHIC COMPILATION

4.31. The topographic compilation may be made on the anharmonic rectifier in much the same manner as that in which the planimetric compilation was made.

4.32. If all the fair-drawing pulls are to contain the contours, their preparation must be delayed until the topo compilation has been completed and checked. When it is essential, the delay may be avoided, since the contour compilation contains nothing closely in register with the detail. Note that, in operation 2.3, that part of the water which affected the detail was included in the planimetric compilation.

4.33. Remove the tabs from the compilation assembly, and proceed to add the topographic compilation. This will complete the drainage pattern and add the contours.

4.34. For reasons of registration, the spot heights which appear on the finished map appear on the black, or detail plate. Ordinarily ground control spot heights will so appear—while photogrammetric spot heights are omitted. For checking purposes, all spot heights are to appear on the topo compilation. These are transferred by the topographers as part of the compilation operation—the transfer is usually by eye. All spot heights, and their elevations in black ink—underline those which are to appear on the published map.

4.35. Assemble the compilation and have a 'matt side down' ozalid made. On the ozalid the compilation is to be completely checked, in particular under the following heads:

4.351. Check edge join with compilation of each adjoining sheet.

4.352. Check registration of water and contours. Sharp draws which might cause trouble to register will be conventionally widened as far as it is possible to do so without destroying character. This is to be done now, by the topographer—not later during fair-drawing.

4.353. Check all spot heights against original data for value and position.

4.354. Check that contours are in sympathy with all spot heights.

4.355. Check that cuts and fills are all shown, and that they are consistent with the topography.

4.356. Check that road shapes are consistent with the topography; investigate and correct discrepancies. Check road and railway grades.

4.357. Label all contours, siting as well as possible in open spaces.

4.358. Officer check of compilation as a whole.

4.359. Any alterations arising from the above checks will be shown on the ozalid in black ink. Trace alterations and amendments to kodatrace compilation, in pencil, using light table. Spot heights, position and elevation, to be in black ink.

4.360. Cuts and fills having been checked, ink these in, their outlines exactly as they are to appear on the finished map.

4.361. The following are in ink on the finally checked compilation: ground spot heights and elevations, underlined; other spot heights, not underlined; cuts and fills.

4.362. Pass to camera for enlargement.

5. CAMERA AND HELIO, PREPARATION OF BLUES

5.1. FAIR-DRAWING MEDIUM AND SCALE

5.11. Fair-drawing is done on enamelled paper, and the blues are made on this. This paper appears to possess all the advantages of enamelled zinc, without the disadvantages the latter has in certain applications. Waterproof, it is substantially distortion free, and the special ferro-prussiate process was designed to take advantage of this.

5.12. Enlargement of pencil compilation is effected by projection to topo base kodaline, x and y distortion coefficients of which are nearly equal.

With publication at 25,000 fair-drawing may be at 20,000, at which scale a sheet 10 × 15 km. is 50 × 75 cm. In making this enlargement exact scale is not important, since the finished fair-drawing is to be rephotographed to exact scale. What is important is that the enlargement be with the easel normal to the optical axis of the camera as, otherwise, trouble will be experienced when rephotographing to publication scale. Squaring up must therefore be precise.

5.2. FERRO-PRUSSIATES

By use of the ferro-prussiate solution, mixed and applied as under, blue line impressions on a white background are obtained by contact in a printing frame, and without the use of a press. The ferro-prussiate is rather simpler, and is quicker, than the bichromate stain system.

5.21. Mix ferro-prussiate solutions A and B as under; keep in opaque containers.

Solution A: 3 oz. potassium ferro-cyanide, grind to powder, dissolve in 20 oz. water.

Solution B: 7 oz. ferric ammonium citrate (green scales) dissolved in 20 oz. water.

5.211. Preparatory to coating a sheet mix:

    1 oz. solution A,
    1 oz. solution B,
    2 oz. water,
    4 oz. industrial spirits.

5.212. Lay a sheet of enamel paper on a smooth surface and pour on about a teaspoonful of mixed solution. Add pumice. With cotton-wool rub the solution and pumice into the paper, rubbing in small circles. Continue for about five minutes covering the sheet as evenly as possible, adding pumice and solution as necessary. Carefully remove the pumice and apply a thin even coat of the solution (less pumice) to the whole area. Drying may be done in the whirler. When dry, the sheet may be exposed. The coating may be done in ordinary daylight, or artificial light, but direct rays of the sun should be avoided.

5.213. Normal exposure is about the same as for gum reversal, say 6–10 min. to a single arc at $4\frac{1}{2}$ ft.

5.214. Develop by wetting with water and wiping lightly with cotton-wool until the image is clean and the surplus colour washed away. Better blues usually result from the addition of a little hydrochloric acid to the developing water.

5.215. When the impression is required immediately, squeegee and dry in a drier. Owing to the enamel, the paper is impervious to water and pulls from the same negative, made at the same time, will be of the same size.

5.216. Where the contours are placed on the compilation after the detail has been enlarged, operation 5.11 above must be to the exact size of the negative obtained from the detail alone. Where the consideration of elapsed time is not important it is better practice to delay enlargement until the contours have been added.

5.217. Spoils, unused spares, or the sheet from which the blue was traced on a transparency, may be cleaned off with ammonia, and resensitized.

6. FAIR-DRAWING

The fair-drawing system and organization, with related camera and helio processes, has been designed with several ends in view:

> To reduce the elapsed hours per sheet.
> To reduce the man hours per sheet.
> To improve the quality of the work.
> To reduce proof-reading.
> To reduce photo-writing, and number of proofs.
> To eliminate litho draughting work.

Most of these factors are related. The less time the fair-drawing has to become dirty, the better copy it makes in all respects. Elapsed time is reduced by cutting the detail sheet into a number of pieces, and working one man on each piece. Man hours have been reduced by means of the organization, and because of the increase in skill which results from specialization.

The enamel surface is such that high contrast between it and the linework makes for good negatives which require photo-writing for little in addition to duffing, and the drawing does not get time to get dirty from days and perhaps weeks of handling in the draughting room. Checks on previous stages reduce the work of the proof-readers, and consequently of the photo-writers. The use of the interpreted, marked-up, photograph as the authority for detail simplifies the work of the draughtsman by keeping always before him an actual photograph of what he is mapping.

Hand lettering is not used; letterpress is more rapid and, by reducing the draughting operation to one of merely sticking down names, removes a further opportunity for the work to become soiled and the linework scrubbed.

When work on a large number of sheets is in hand, pulls being ready first thing in the morning, the job should go to the camera the following day at the latest. This maintains interest in that the workman sees immediately the fruits of his labours, with consequent improvement in morale. There is also a valuable training opportunity, since a man can be shown, while the work is still fresh in his mind, where and how his work can be improved to decrease that of the photo-writers and better the job. The aim is that the photo-writer shall never be required to clean out poor quality linework.

Fig. 81. Fair-drawing organization chart.

SUBDIVISIONS OF FAIR-DRAWING OPERATION

6 FAIR-DRAWING

6.1 DETAIL PLATE (black).
Drawn in several pieces on enamel paper.

6.110 WORK ON UNCUT BLUE.
6.111 Determine position of sheet cuts. Cuts must be on the grid line.
6.112 Ink grid except on cut lines.
6.113 Draw railways, whole area, and surveyed main roads.
6.114 Draw internal edges across cut lines.
6.115 Tape pull to mounting sheet, and cut.

6.120 DETAIL, WORK ON EACH PIECE.
6.121 Draw roads and paths.
6.122 Cut out other roads and footpaths, pecked lines.
6.123 Houses, powerlines, churches.
6.124 Bridges.
6.125 Cuts, fills, quarries.
6.126 All other detail.
6.127 Draw or transfer vegetation stipple.

6.14 ASSEMBLY, detail plate.

6.141 Coat mounting sheet with rubber cement.
6.142 Mount each piece as received.
6.143 Mount place names, siting as nearly clear of detail as possible.
6.144 Mount all other material; spot heights, road classification, p.p.'s, p.p. numbers, trigs, etc.
6.145 Mount all marginal data.
6.146 Proof-read assembled drawing, using the checking form.
6.147 When operations 6.29 and 6.40 complete, pass to camera.

NOTE. Procedure of 6.14 applies where the major portion of the map is the black plate. In country where the black, or detail, is but a small proportion of the sheet—which would consist then chiefly of contours and the drainage pattern—the procedure is modified accordingly.

6.130 NAMES AND MARGINAL DATA, letterpress requirements to be passed to typographer.
6.131 List place names.
6.132 List spot heights.
6.133 Sheet number and title, compilation note.
6.134 Grid and geographical numbers.
6.135 List photo numbers.
6.136 Index box.
6.137 Reliability box.
6.138 Convergence and declination.
6.139 Measure grid on pull, construct scale.

6.2 WATER PLATE (blue).
Drawn on one piece on matt acetate.

6.21 Tape sheet of matt acetate to (topo) pull and establish pencil corner marks.
6.22 Draw water outlines in red poster ink and all that is to appear on finished map. Lakes are solid (for stipple).
6.23 Mount place on assembled detail plate in exact registration.
6.24 Cut out for bridges.
6.25 Cut out for intermittent streams, if shown as such.
6.26 Check for tight registration to detail, e.g. ditches along roads, at 'works of man',
6.27 Blue may run through black names—no names are shown in blue, add blue grid numbers to appear within sheet. These may overprint detail provided legibility of neither is compromised.
6.28 Add parts of surround which are to appear in blue.
6.29 Ink in registration marks to their exact position with regard to black grid corners, erasing the pencil marks of 6.21 above.

NOTE. Authority for all colours is marked-up photograph.

6.3 CONTOUR PLATE (brown).
Drawn on (topo) enamel pull.

6.31 Draw contours, paying particular care to line gauge.
6.32 The use of a contour pen is to be avoided as tending to smooth significant small irregularities thus destroying character.
6.33 Draw any sand stipple.
6.34 Prepare trace, on tracing paper, showing contour number positions—cut small windows.
6.35 Place on assembled black, revise position where numbers would overprint on black. Only adjustment should be for names. At same time locate any brown required for legend.
6.36 Replace on brown, mark new positions.
6.37 Mount numbers, add brown for legend.
6.38 Place water transparency on pull in registration. Check all blue-brown registration. Amendment should be in brown.
6.39 Place blue again on brown, in exact registration. Prick through blue corner marks to brown.
6.40 Ink these corner marks carefully—pull ready for camera.

The system requires no drawing or correction whatsoever on the zinc.

The operations are shown in detail, for all plates, on the fair-drawing organization chart, fig. 81.

6.1, 6.3. Paper must be well pounced, and all surplus removed, before work is commenced.

6.111. Sheet cuts are on grid line. Photo-writers scribe the position indicated by the join, which shows on the negative. Occasionally it may be advisable to cut other than on a grid line.

6.113. Railways and engineered roads should be drawn over the whole area, largely for appearance sake, before cutting. Other detail is drawn only for a short distance either side of the proposed cut, to preserve exact continuity.

6.115. Tape the pull down to a sheet of enamel paper, cut deeply enough to mark the under sheet, thus forming guide lines for assembly. Fill in the lines with chinagraph. Number each piece and the corresponding areas of the mount sheet. Number of pieces may be anything from twenty or more, to four. Factors are detail density, time available or allowed, number of men available.

6.122. Broken or packed lines will be first drawn solid. They are broken with a photo-writer's tool. Very sharp clean 'ends' are so obtained. Similarly, fence lines, etc., at a corner are drawn to cross, and the excess removed. Small houses are squared with the same tool.

6.127. Large stipple areas will be drawn, or transferred, after assembly if on either side of a join.

6.13. Make all items, except 6.139, ready in advance.

6.140. The pieces are fastened down with rubber cement, mounted in the exact position indicated by the cuts from 6.115.

6.143. Mount names parallel to a grid line, to read from the South or East as is suitable. This rule is followed even where the sheet lines are at an angle to the grid. Place so that least amount of detail is covered. Make detail good to edge of type, if necessary.

6.146. Omissions and alterations may be put right at this stage more quickly than when picked up by the proof-readers and passed on to the photo-writers. Use PR check Form 1. Linework will not be strengthened or gone over at this time. It should be done correctly in the first instance. Note faulty linework and later show this, the negative, and the proof, to the man responsible so that he will learn to recognize what will and will not reproduce. It is a good thing for draughtsmen to have had some photo-writing experience in order that they may appreciate what may have to be done.

6.4. Pens, drawing pens, and all tools are to be kept *scrupulously* clean at all times, and blade pens will be properly sharpened. *Note.* Pen wear on enamel paper is slight—it is not like zinc or foil. Drawing pens need not be very sharp.

All pens will be wiped dry if out of use for even a few seconds. A layer of dried ink must not be allowed to form. Ink will be stoppered at all times except when the stopper is removed for the purpose of filling a pen. Pens will not be dipped in India ink; fill with a quill. Good linework cannot be done with thickened, dusty ink in a dirty pen.

The finest line, if it consists of a solid opaque ridge of ink, will reproduce, but even a wide line is liable to appear weak in the negative if there is a variation in the density, that is to say, in the thickness, of the layer of ink deposited on the drawing medium. The ink must look wet coming off the pen to form the line. The pen must contain sufficient ink. As the pen becomes nearly dry the layer of ink deposited thins, although the width of the line will not change noticeably. Learn to recognize the characteristic change in appearance as pen becomes nearly dry. Re-ink at this time.

6.5. Avoid passing the hand or arm over completed linework. Work towards you. It is as well to cover the lower portion with a sheet of clean litho stock. The ink will not hold if the surface is moistened with perspiration, or becomes greasy. When this happens, repounce.

7. FINAL PHOTOGRAPHY AND FIRST PROOF

7.1. CAMERA WORK

7.11. All precautions are taken that the Kodatrace does not change in size, and the compilation operations being done on separate strips reduces the time during which this may occur. It is difference in x and y distortion coefficients which causes trouble, not equal expansion or contraction on both axes. If, in operation 5.12 (projection of the Kodatrace compilation), it has been possible either to set to exact size, or to leave only radial residuals, no further trouble will be experienced. If this has not been possible in the projection operation, it is still possible to rephotograph to nearly exact size and shape since the tilting of the easel in copying has a different effect than when projecting on to it.

7.12. Kodatrace is not ideal as a compilation medium, but careful photographic work is quite capable of rectifying the small distortions which may be present. In any case grid, control and detail will always be in sympathy.

7.13. The detail having been photographed to scale, the other plates are photographed at the same camera setting. The fair-drawing department has effected registration at the fair-drawing scale, and their registration marks are in sympathy. If therefore all plates are photographed at the same setting, the negatives too will be in register. The responsibility of the camera operator is then to attempt to rectify small distortions, and to ensure that all plates are photographed under identical conditions. If registration marks on each fair-drawing plate are not exactly similar, the photographer must report this to the Officer in charge Fair-drawing—the photography will not be continued until the registration marks are correct.

7.14. Glass plates are used.

7.2. PROOF

7.21. As soon as the plates are dry, and without any photo-writing whatsoever, the glass is passed to the printers for three-colour proofs.

7.22. About a dozen proofs are required, and they should be in exact registration. Again, the glass registration marks are supposed to be to size, and (in a hand press) if two marks are on, the others should be. In case they are not, printing will be carried out at the best internal fit, and the discrepancy at the corner marks will be noted by the proof-readers and its cause traced down.

7.23. Brown-blue proofs are usually required in addition.

7.24. As each exposure is completed, pass the glass to the photo-writers so that they can get on with the duffing.

7.25. Pass completed proofs to proof-reading department.

7.3. Proof-reading and Photo-writing

7.31. The printer is the first man who sees, and the proof-reader the first man who works upon, the completed job which, heretofore, has had no existence as an entity. If the work has been well done the colours will fit one another and, what is of immeasurably greater importance, the map will fit, and truly depict, the ground. It is a part of the proof-reading job to maintain a record of the number and kind of corrections necessary in order that action may be taken to reduce or eliminate certain kinds of error.

7.32. Final responsibility that the map is in agreement with the basic data rests with the proof-readers, except with respect to photogrammetric accuracy. Basic data are usually as under:

Planimetry, marked-up photos—occasionally library photos. Interpreted photographs will be accepted, but any questionable or illogical interpretation will be referred to Officer in charge Interpretation—whose decision is *final*.

Road and railway classification. Base-map, intelligence trace, or as directed.

Topography—topographic compilation. Policy of (i) above will be followed.

Place names and spelling, base-map.

Edge joins, latest available data.

Geographical and convergence data, trig positions. As supplied by control department.

Other marginal data, as supplied by Officer in charge Work, and/or the specification.

7.33. A corrected proof for each colour is supplied to the photo-writers. The detail proof is ordinarily divided into a number of pieces. Before reading is commenced, the map or piece is to be mounted on a piece of white stock size to leave a good margin on all sides. Legible writing, brief clear instructions, accurate keying to the exact map spot to which reference is made, are essential.

7.34. Corrected proof for a particular colour having been passed to the photo-writers, and corrections reported as made, it is a further responsibility of the proof-reader to check that the corrections are in fact made, and made correctly. Upon the proof-reader's O.K., the map will be printed and distributed.

7.35. When a second proof is required, it will be subjected to such further check as may be necessary.

7.36. Actual sequence of proof-reading operations in detail will be made in accordance with PR Forms 2 to 6.

SURROUND CHECK TO FAIR-DRAWING

PR FORM 1

Job No. ....   Job Sheet No. ....
Publication No...................   Sheet Name ....
Index No. ....   Date ....

ITEM   INITIALS

1	Scale	
2	Sheet name	
3	Edition and date	
4	Sheet number	
5	Publication number	
6	Company name	
7	List of verticals	
8	List of obliques	
9	Reliability diagram	
10	Footnote all legible and same as specimen layout	
11	Convergence and declination in agreement with 1 : 25,000	
12	Date of declination	
13	Magnetic north arrow	
14	Index incidence diagram: (a) Sheet in correct relative position to 1 : 25,000	
	(b) Co-ords and letter correct (on blue)	
15	Check corner co-ords and run a two-point check against base-map to ensure no boners in establishing grid numbers	
16	Scale; trial check map vs. grid (Scale agrees within permissible error)	
17	Series block—all colours	

NOTE

Supervisor's Signature................................
Date Time Signature ....

EDGE

PR FORM 2

Job No. ....   Job Sheet No. ....
Index No. ....   Time Started ....
Sheet Name ....   Time Finished ....

NOTE. See shift supervisor for correct material to check to.

INITIALS   DATE

I	Continuity of all black detail including wood features:		
	On North edge............		
	„ South „		
	„ East „		
	„ West „		
2	Road and railway classification:		
	On North edge............		
	„ South „		
	„ East „		
	„ West „		

Supervisor's Signature...............................
Date Time Signature ....

SURROUND

PR FORM 3

Job No. ....   Job Sheet No. ....

Index No. ....   Time Started ....

Sheet Name ....   Time Finished ....

NOTE. See shift supervisor for correct material to check to.

		INITIALS	DATE
I	Corner co-ordinates, N.W.......... S.W.......... N.E.......... S.E..........		
2	Are suffixes N.S.E.W. shown and correct on all the foregoing?		
3	Grid numbers, all edges, small number on even 10's eastings and northings		
4	Magnetic north arrow		
5	Declination		
6	Date of declination		
7	Index incidence diagram and letter (Check to officer material supplied)		
8	Sheet number		
9	Publication number and imprint		
10	Edition and date		
11	Notes: Notes applicable to sheet, list of photographs used		
12	Scale; trial, check map scale vs. grid		
	TRIG AND P.P.'s		
I	Scale trigs—6 figures		
2	Compare with trig list		
3	Check that all trigs and churches are included		
4	P.P.'s shown		
5	Position by detail on ozalid		
6	Photo numbers		
7	Flight numbers		
8	The name of this sheet is		
9	The scale of this sheet is		

Supervisor's Signature.............................

Date Time Signature ....

DETAIL

PR FORM 4
Publication No....................
Index No. ....
Sheet Name ....

Block Sheet No. ....

ITEM	BLACK DETAIL	INITIALS	DATE
1	Roads, and classification and shape		
2	Bridges, all places where full rivers cross roads		
3	Dams—named, cuts, fills, cliffs, bridges		
4	Railways, and classification		
5	Stations, locations, symbols and named		
6	Road and railway bridges		
7	Trams, and light railways		
8	Buildings and special features, shape and relative position		
9	Fencelines, hedges, boundaries, walls		
10	Names and heights on features		
11	Trees along roads		
12	Stipple against stipple key		
13	Does anything look odd?—square by square		
	NAMES		
1	All main features named		
2	Spelling correct		
3	Keyed to feature		
4	Check sheet name and spelling		
	SPOT HEIGHTS		
1	Spot heights (a) all included, (b) list of base-map spot heights which are omitted is shown on attached sheet, (c) height is on and legible		

Supervisor's Signature..............................
Date Time Signature ....

CONTOURS

PR FORM 5

Job No.	Job Sheet No. ....
Publication No...................	Portion ....
Sheet Name	Date ....
Index No.	

ITEM		INITIALS
1	Contours keyed to all BLACK detail	
2	BROWN fits BLUE, all but minor corrections on brown	
3	Location of all spot heights	
4	Contours fit spot heights	
5	Contour numbers legible and reading uphill	
6	Contours broken for quarries, cliffs, etc.	
7	Continuity of contours	
8	Depression contours broken with strokes on LOW side of contour	
9	Gradients within limits—roads, railroads and main rivers	
10	Mounds	
11	Edges tied to latest material available	
12	All corrections on transparency	

Supervisor's Signature.............................

Date Time Signature ....

WATER

PR FORM 6

BLUE

Job No. ....   Job Sheet No. ....

Index No. ....   Time Started ....

Sheet Name ....   Time Finished ....

NOTE. See shift supervisor for correct material to check to.

ITEM		INITIALS⌐	DATE
1	Position, solid and broken drainage and marsh		
2	Marine contours		
3	Gradient of drainage		
4	Blue edge ties on latest material available, N.......... S.......... E.......... W..........		
5	Ensure blue is broken for bridges and roads		
6	Ensure direction of flow		
7	Check to latest material for identification of all lakes and ponds		

Supervisor's Signature..............................

Date Time Signature ....

MULTIPLEX AND STEREOPLANIGRAPH: CONSIDERATIONS GOVERNING MINIMUM CONTOUR INTERVAL

Refinements of optical design and of the calibration and matching of projector lenses have resulted in improved performance of multiplex equipment.

It has been explained that the optical train of the multiplex consists of three parts, air camera, reduction printer, and projectors. The design and calibration of the equipment are directed to obtaining, as nearly as may be, a distortion-free image. This means that the distortions arising in each unit should nicely balance out, and further implies accurate matching of the projectors, one to another.

If this end is attained (within readable limits) there are uncompensated only the following: variation in distortion characteristics of lenses of the same family, and distortions due to differential film shrinkage.

Air-camera lenses which differ by more than the allowable tolerance from the standard distortion for which the reduction printer was designed may be rejected, at least when working at H/1000.

This leaves differential film distortion as the bugbear.

Good operators can read sharp multiplex models to plus or minus 0·03 mm. or better. Now, differential distortion of 0·008 mm. on the 6 in. negative plane, at 30° on the base-line, will cause a heighting difference of this amount. It is apparent therefore that the instrument can reflect differential film distortions of magnitudes which do occur.

It is equally apparent that it can also reflect differences in distortion of air lenses of the same family.

How closely the optical design and calibration compensates the various lens distortions, quite frankly the writer does not know. What can, however, easily be checked is how closely the model spot heights agree with corresponding ground elevations. That is to say the combined effect of residual optical distortions, plus film distortion, can readily be measured. The writer has conducted numerous tests of this nature upon well-known makes of equipment and has found the arithmetic mean deviation to be about 1/4000 of the flight altitude. This means that the actual mean precision of the equipment is about one-tenth of a millimetre at model scale.

What proportion of this is due to optical, and what to film, distortion cannot be estimated from these tests. The effects could, however, be separated by repeating the tests using glass plates.

The tests were with low-altitude photography, but the data may fairly be assumed to apply to higher altitudes. The processing procedure followed standard good practice, and the photography was good. The operators had model acuity of about 0·02 mm., but this is not unusual.

The whole of the foregoing leads to the conclusion that *from the point of view of heighting accuracy alone* there is no case for equipment of higher precision than the multiplex unless glass plates are used.

But heighting accuracy as such is not the only consideration. In steep terrain a flight altitude of 1000 times the interval gives a multiplex scale so small that contours at the required interval cannot be drawn. With a sidehill slope of 20°, H/1000 contours are 1 mm. apart (arctan 0·36/1 = 20°). On slopes much steeper than this the error of the tracing table pencil in drawing the contour line makes the H/1000 interval impracticable.

Accordingly flight altitude must be decreased with consequent increase in cost both of mapping and of photography.

It is suggested that this, rather than any consideration of map scale, is the actual factor limiting the economic application of the multiplex. Publication scale larger than multiplex manuscript scale is practicable, but only in the circumstance that (as explained on p. 145) the resultant planimetric precision is consistent with map use.

The stereoplanigraph type instruments of Zeiss, Wild, and Pouvillier, have important advantages in such circumstances. These are: larger 'natural' scale, that is larger instrumental air-base; magnification in the model viewing optics; and means to produce the pencil manuscript at a variable scale larger than the natural scale of the instrument.

The new Kelsh instrument, with larger projectors and longer instrumental base possesses most of the advantages above and could thus be expected to work at H/1000 or thereabouts in steeper terrain than the multiplex.

To recapitulate, the multiplex can be relied upon for contour intervals of H/1000 where the ground is comparatively flat. If we adopt quite arbitrarily the rule that the *horizontal* interval should not be less than 1 mm. at model scale we arrive at the following:

Slopes 20° or flatter,   VI H/1000 = 0·36 mm.
Slopes to 31°,   VI H/600 = 0·60 mm.
Slopes to 45°,   VI H/360 = 1·00 mm.

This may serve to reconcile the apparently gravely complicating statements made from time to time by different authorities. The table is given by way of illustration only.

On the other hand, the stereoplanigraph instruments can work at H/1000 or better under almost any ground conditions. Using glass plates (and of course a matched air camera lens) such instruments are capable of much higher precision than this.

THE KELSH PLOTTER

DESCRIPTION. The Kelsh plotter is a projection-type instrument employing contact diapositives.

A Kelsh model differs in appearance from a multiplex model at the same scale only in that the definition, and the illumination, of the former tend to be better. Detail and topography are taken off by means of a multiplex tracing table. Setting up and operation do not require further explanation.

It is to be noted that, since it is a two-projector instrument, it is not necessary to provide six degrees of freedom to each projector, and accordingly controls are simplified. The model having been set up and scaled, is horizontalized by the bar which may be moved on a three-point system either independently of, or with, the frame footscrews.

Model scale is variable between $4\frac{1}{2}$ and $5\frac{1}{2}$ times contact scale. Data of p. 127 could be taken as referring to a Kelsh working 4 times enlargement from 6-in. 9 by 9 contact diapositives.

In the 6-in. Kelsh as manufactured by The Instruments Corporation, PD is variable from 27 in. to 33 in.—700 mm. to 850 mm. Corresponding instrumental air base variation is 16 in. to 20 in.—400 mm. to 500 mm.—for 60% forward lap.

The projectors may readily be adapted for $8\frac{1}{4}$ in. and other lenses.

KELSH SYSTEM OF ILLUMINATION. The problem of providing illumination for a full-size diapositive has been mentioned briefly, p. 127. Von Gruber(1), p. 304, points out that, compared to an image viewed by transmitted light, in projection upon a screen only the *one twenty-thousandth part* of the available light may reach the eye.

It is understandable, therefore, why previous designers of instruments employing contact scale transparencies rejected simple solution by means of direct optical projection. In spite of its other manifest advantages, the illumination problem appeared to be insurmountable. In the fine instruments of Zeiss, Wild, Pouvillier, the image is viewed by transmitted light, and spatial intersection recovered by optical and mechanical linkage of great ingenuity, and manufactured necessarily to almost superhuman limits of precision, that the space rods may accurately parallel the paths of the two corresponding rays of light.

In a full-size projector, it is evident that illumination is required, not of the whole format, but only of that part which is being examined—the area of the tracing table platen.

This is what Kelsh does, confines the illumination to the needed area, a spot of a square inch or so on each diapositive, rather than attempting to light the whole 80 square inches of format.

The illumination means may be likened to a flashlight shining on each diapositive. As the tracing table moves over the model, the flashlights automatically move in sympathy, being continuously directed to the proper area in each diapositive.

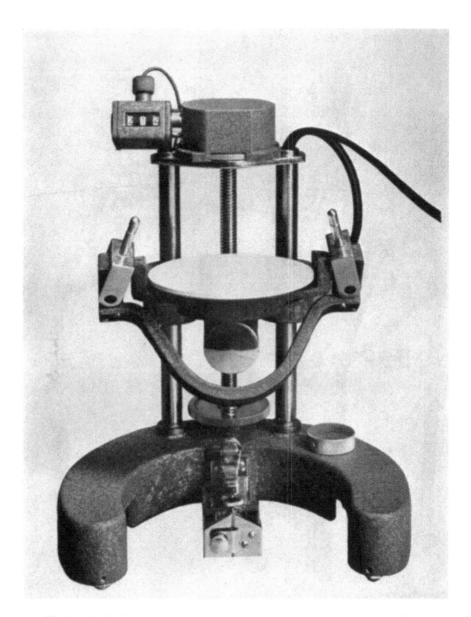

Fig. 82. The Kelsh plotter, as manufactured by the Instruments Corporation, Baltimore, Maryland. (Photograph by permission of the Instruments Corporation.)

To face p. 174

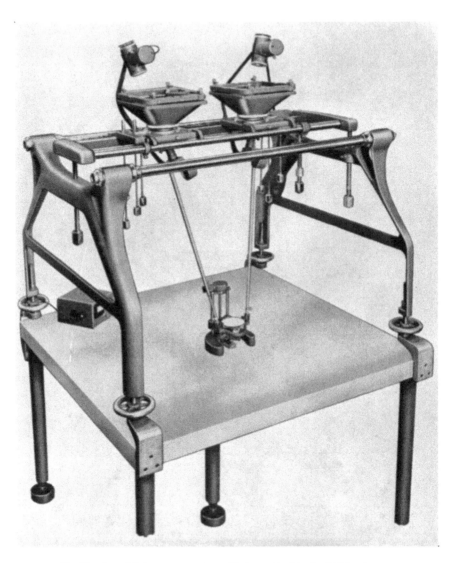

Fig. 83. A multiplex type tracing table is used with the Kelsh plotter.
(Photograph by permission of the Instruments Corporation.)

To effect this, each lamp housing is linked to the tracing table. As the table moves, following contour or detail across the model, the linkage maintains each lamp always in proper position and alignment to point towards the centre of the tracing table, along a line through the centre of the projector lens.

It will be understood that the linkage need not be particularly precise to effect the purpose, accordingly construction is light and simple, as may be seen in examining fig. 82. Fig. 83 shows the tracing table itself.

The light source may be a standard projection lamp, or alternatively a Western Union point source tube, with simple condenser system. In the multiplex some 400 square inches of projected area are illuminated, while the Kelsh system illuminates but 10 or 12 square inches.

Thus we have a projection instrument using diapositives of the full size of the taking camera. The illumination problem is solved and, with it, attendant problems of definition, and of depth of focus, in the projected image.

MATCHING OF PROJECTOR AND CAMERA LENSES. Whatever the method of reprojection, whether optical and mechanical as in the earlier fine instruments, or by direct reprojection in the multiplex and the Kelsh, the optical elements of projection must match the optics of the taking camera.

In air lenses of the best makes, differences in distortion characteristics are apparent in lenses of the same family. Further, the 6-in. wide angle survey lens, say, of Ross has characteristics which differ from those of the metrogon by Bausch and Lomb. Previously these differences between individual lenses of the same make, and between similar lenses of a different make, provided an operational problem the economic solution of which presented serious difficulty.

Pouvillier, and others, use a series of meticulously calibrated goniometer lenses; from these may be chosen pairs which are a good match to a particular air camera lens.

KELSH COMPENSATION. In the Kelsh instrument, integral with the linkage to the tracing table is a spherical cam. This cam is ground to such a pattern, and arranged in such a manner, that it imparts a slight axial motion to the projector lens. according to the angular zone in which the table is plotting. This motion is designed to compensate the known difference in distortion between the taking lens and each projector lens.

The distortion* curve of a particular air lens being known from its calibration data, a cam may readily be ground exactly to compensate the difference between this curve and that of the particular projector lens in use.

In this way, it is a simple matter to compensate any individual Kelsh projector unit to a particular air camera regardless of make, or of family variation from the norm.

---

* In the table of notation, lens distortion has, at least by implication, been defined as arising from inaccuracies. At this stage it may be well to amplify this. There is a difference in position between an image point as projected by a survey lens and the true perspective position of that point. Such differences are expected, are known, and arise from other lens design considerations. They are radial.

This radial component may be thought of as a displacement, whatever we choose actually to call it.

Tangential components are also found. They arise, for the most part, from minute centering errors in lens components. Some physicists prefer to call the radial component a displace-

PRECISION OF THE KELSH PLOTTER.   The instrument can thus compensate all primary sources of optical error except tangential distortion. Working from a camera using glass plates, error from dimensional instability of the emulsion base does not arise.

No test data are known to the writer of the precision attained by an accurately compensated Kelsh plotter working from glass negatives.

The analysis above, however, leads to the conclusion that one should expect bare ground spot height precision of the order of H/6000 from optimum glass plate photograph—perhaps better. This is equivalent to saying that Kelsh model readings of a tenth of a millimeter should be significant.

Where the Kelsh is working from film, precision is limited by dimensional stability of the base. As any experienced worker in the field knows, air film exposed and processed under operational conditions can rarely be depended upon for operation at C factor higher than 1000.

ment, and the tangential component a distortion.   Such a conception clarifies lens performance—the displacement is more or less deliberate, and can be designed out in a complementary part of the system.   Even variations in its magnitude can be controlled as we have seen.   On the other hand, the tangential component, the 'distortion', can be controlled only by manufacturing and assembling refinements.   In multiplex projector lenses where, because of the small physical size of the assembly, extreme precautions must be taken, tangential distortion causes a disconcerting series of phenomena, particularly in long bridges.

INDEX

Printed in the United States
By Bookmasters